SYMPOSIA OF THE ZOOLOGICAL SOCIETY OF LONDON NUMBER 60

Reproduction and Disease in Captive and Wild Animals

SYMPOSIA OF THE ZOOLOGICAL SOCIETY OF LONDON NUMBER 60

Reproduction and Disease in Captive and Wild Animals

To mark the 50th anniversary of the Wellcome Trust

Edited by G. R. SMITH

and J. P. HEARN

Institute of Zoology, The Zoological Society of London, Regent's Park, London, England

Published for THE ZOOLOGICAL SOCIETY OF LONDON

by CLARENDON PRESS · OXFORD

1988

Oxford University Press, Walton Street, Oxford OX2 6DP

Oxford New York Toronto
Delhi Bombay Calcutta Madras Karachi
Petaling Jaya Singapore Hong Kong Tokyo
Nairobi Dar es Salaam Cape Town
Melbourne Auckland

and associated companies in
Berlin Ibadan

Oxford is a trade mark of Oxford University Press

Published in the United States
by Oxford University Press, New York

British Library Cataloguing in Publication Data
Reproduction and disease in captive and wild animals.
1. Animals. Reproduction. 2. Animals. Diseases
I. Smith, G. R. (Geoffrey Robin), 1929–
II. Hearn, J. P. (John P.), 1943–
III. Zoological Society of London
IV. Series
591.1'6
ISBN 0–19–854007–8

Library of Congress Cataloging in Publication Data
Reproduction and disease in captive and wild animals.
(Symposia of the Zoological Society of London; no. 60)
"To mark the 50th anniversary of the Wellcome Trust."
Includes bibliographies.
1. Wildlife conservation. 2. Reproduction. 3. Animals—Diseases. 4. Wellcome Trust (London, England) I. Smith, G. R. (Geoffrey R.) II. Hearn, J. P. III. Wellcome Trust (London, England) IV. Title. V. Series.
QL82.R47 1988 639.9 88–15250
ISBN 0–19–854007–8

Set by Promenade Graphics Ltd., Cheltenham
Printed in Great Britain by
The Alden Press, Oxford

Contents

The hormonal and hypothalamic control of primate sexual behaviour

A. F. DIXSON & S. A. C. LLOYD

Embryo manipulation for the regulation of reproduction and disease

P. M. SUMMERS & J. P. HEARN

Diagnosis of pregnancy and disease by ultrasound in exotic species

G. H. du BOULAY & O. L. WILSON

Immunoassays with special reference to ELISA methods

A. VOLLER & D. E. BIDWELL

Anaerobic bacteria as pathogens in wild and captive animals

G. R. SMITH

Organizing Committee

J. P. HEARN, MRC/AFRC Comparative Physiology Group, Institute of Zoology, The Zoological Society of London, Regent's Park, London NW1 4RY, UK.

R. D. MARTIN, Anthropologisches Institut, Universität Zürich-Irchel, Winterthurerstrasse 190, 8057 Zürich, Switzerland.

W. PLOWRIGHT, Whitehill Lodge, Goring-on-Thames, Reading, Berkshire RG8 0LL, UK.

K. B. SINCLAIR, The Wellcome Trust, 1 Park Square West, London NW1 4LJ, UK.

G. R. SMITH, Nuffield Laboratories of Comparative Medicine, Institute of Zoology, The Zoological Society of London, Regent's Park, London NW1 4RY, UK.

Contributors

ABBOTT, D. H., MRC/AFRC Comparative Physiology Group, Institute of Zoology, The Zoological Society of London, Regent's Park, London NW1 4RY, UK.

BIDWELL, D. E., Nuffield Laboratories of Comparative Medicine, Institute of Zoology, The Zoological Society of London, Regent's Park, London NW1 4RY, UK.

DU BOULAY, G. H., Nuffield Laboratories of Comparative Medicine, Institute of Zoology, The Zoological Society of London, Regent's Park, London NW1 4RY, UK.

DIXSON, A. F., MRC Reproductive Biology Unit, Centre for Reproductive Biology, 37 Chalmers Street, Edinburgh EH3 9EW, UK.

HEARN, J. P., MRC/AFRC Comparative Physiology Group, Institute of Zoology, The Zoological Society of London, Regent's Park, London NW1 4RY, UK.

HOLLOWAY, A. J., MRC/AFRC Comparative Physiology Group, Institute of Zoology, The Zoological Society of London, Regent's Park, London NW1 4RY, UK.

LAWS, R. M., British Antarctic Survey, NERC, Madingley Road, Cambridge CB3 0ET, UK.

LLOYD, S. A. C., MRC Reproductive Biology Unit, Centre for Reproductive Biology, 37 Chalmers Street, Edinburgh EH3 9EW, UK.

MACLARNON, A. M., Department of Anthropology, University College London, Gower Street, London WC1E 6BT, UK.

MARTIN, R. D., Anthropologisches Institut, Universität Zürich-Irchel, Winterthurerstrasse 190, 8057 Zürich, Switzerland.

MOORE, H. D. M., MRC/AFRC Comparative Physiology Group, Institute of Zoology, The Zoological Society of London, Regent's Park, London NW1 4RY, UK.

PLOWRIGHT, W., Whitehill Lodge, Goring-on-Thames, Reading, Berkshire RG8 0LL, UK.

SMITH, G. R., Nuffield Laboratories of Comparative Medicine, Institute of Zoology, The Zoological Society of London, Regent's Park, London NW1 4RY, UK.

SUMMERS, P. M., MRC/AFRC Comparative Physiology Group, Institute of Zoology, The Zoological Society of London, Regent's Park, London NW1 4RY, UK.

VOLLER, A., Nuffield Laboratories of Comparative Medicine, Institute of Zoology, The Zoological Society of London, Regent's Park, London NW1 4RY, UK.

WILSON, O. L., Nuffield Laboratories of Comparative Medicine, Institute of Zoology, The Zoological Society of London, Regent's Park, London NW1 4RY, UK.

Symp. zool. Soc. Lond. (1988) No. 60 1–5

Introduction

R. M. LAWS[1]

British Antarctic Survey
NERC
Madingley Road
Cambridge CB3 0ET

Over the next half century we can expect a doubling of the human population, with consequent reductions in the space available for wild animals. In addition, there is likely to be a fragmentation of animal populations into smaller units—islands in the sea of humanity—with the prospects for their survival depending on more interventive management. Development of a new field of biology is required now if we are to cope with these problems. The urgent priorities are for a greater understanding of animal ecology, nutrition, reproduction, genetics, disease and veterinary care to provide the necessary scientific basis for conservation. If the knowledge is not acquired and the necessary management action not identified and implemented, we can only predict a future for this planet marked by catastrophic losses of plant and animal diversity.

In introducing the first theme, reproduction, I would like to mention two examples of studies in reproductive biology from the Antarctic and Africa that contributed to conservation and management. The first was the work I carried out between 1948 and 1953 on the southern elephant seal, *Mirounga leonina*, where estimates of age at maturity and pregnancy rate were important in constructing what was one of the first population models for a large mammal (Laws 1960). From this model conclusions on a revision of the management regulations for the sealing industry were implemented. A method of monitoring the age structure of the catch (limited to adult males) was proposed and predictions were made about the expected change in the age of the catch under the new regulations. Over a period of 10 years the mean age increased to the predicted level and stabilized there, indicating that a sustainable yield was being obtained (Laws 1979).

The second example relates to the African elephant, *Loxodonta africana*, where studies associated with large-scale culling operations illustrated the great plasticity of reproductive processes under different environmental conditions—often induced by the animals themselves. The mean age at sexual maturity in six populations ranged from 11 years in a productive population in undamaged habitat, to 23 years in a declining population living in

[1] Present address: Secretary, The Zoological Society of London, Regent's Park, London NW1 4RY, UK.

ZOOLOGICAL SYMPOSIUM No. 60
ISBN 0-19-854007-8

a habitat converted by the elephants themselves from woodland and bush to open grassland. The extremes of the range of mean calving intervals were respectively three and nine years (Laws, Parker & Johnstone 1975). Without such studies this great habitat-dependent flexibility of response would not have been appreciated.

In addition to such traditional field studies, the new technologies being applied to reproduction can play some part in preserving diversity and accelerating the breeding of endangered species. Cryopreservation of sperm and embryos, as yet only achieved in very few species, should eventually allow the transfer of genetic material between small captive or wild populations. New non-invasive methods, allowing the early diagnosis of pregnancy, and the assessment of breeding status and potential, can contribute to the implementation of conservation policies. A greater understanding of the dynamics of, and ecological constraints on, small populations is essential. As yet we have detailed knowledge of far too few of these limiting factors on far too few species. Nor should we forget the purely academic and theoretical interest in these subjects.

The papers in this volume address these topics in presenting the results of research on wild and captive animals.

Concerning the parallel theme, there is a steadily growing awareness that infectious diseases can exert significant effects in regulating animal populations. In some instances mortality may occur on a large scale and, when a species is already rare, disease can have a particularly serious, even decisive, effect on its survival and on genetic diversity. Animal populations may also be influenced indirectly by disease. Thus, attempts to reduce their size may be made when they are found to be acting as reservoirs of infections transmissible to man or his domesticated animals. Current controversy over tuberculosis in badgers and cattle is a case in point, as is malignant catarrh in wildebeest and its transmission to cattle.

The subject of wildlife disease offers a great intellectual challenge to the research workers. Such research also offers the prospect of acquiring new knowledge applicable to man and domestic animals as well as to conservation. These are not, however, the only considerations. Our duty towards the welfare of farm, companion and laboratory animals is clear; we should be equally concerned to promote research designed to reduce the transmission of disease from domestic animals to wildlife. In a few parts of the world, for example North America, this responsibility is adequately discharged, but a much greater effort is needed elsewhere, particularly in tropical Africa.

Some of these considerations are dealt with by the authors in this volume. Their contributions, which embrace infections caused by viruses and bacteria, illustrate the benefits to be derived from the study of disease in both free-living and captive wild animals.

To exemplify the kinds of problems that need to be tackled, I would like to mention two interesting examples of studies of infectious diseases in wild animals that I have been involved in myself. Again, the first relates to the Antarctic, the second to Africa.

In the Antarctic spring of 1955 a sledging party discovered some 3000 crabeater seals, *Lobodon carcinophagus*, on the sea ice of Prince Gustav Channel near the tip of the Antarctic Peninsula (Laws & Taylor 1957). In the main concentrations of some 2500 seals, deaths were first seen on 4 September; 66% were dead by 29 September and the average mortality was 85% by 21 October. This is the normal pupping season, but there were no live births, although there were numerous abortions. Carcasses were examined and specimens were collected for later laboratory examination. Unfortunately, conditions and lack of expertise and fixatives meant that detailed and critical post-mortem and histological examination was not possible. It was concluded that death was due to disease, probably a virus infection. Mortality was random with respect to age. The disease did not spread to some 300 Weddell seals, *Leptonychotes weddellii,* in the area.

Between December 1979 and October 1980 more than 400 harbour seals, *Phoca vitulina*, died along the New England coast of acute pneumonia associated with influenza A virus. Carcasses were examined by specialists. The mortality rate was much lower, but the striking similarities between the harbour seal epizootic and the crabeater seal and other large-scale mortalities in the past 'suggest that these events may be linked by common biological environmental factors that we are only beginning to recognize' (Geraci *et al.* 1982). Later material from harbour seal deaths from June 1982 to March 1983 led to the discovery that seal viruses were most closely related to avian virus strains, quite distinct from mammalian viruses, including those from the earlier harbour seal epizootic. This association suggests that transmission of avian viruses to seals may be a step in the evolution of influenza viruses in nature, or that avian viruses may be frequently transmitted to seals and associated with disease outbreaks (Hinshaw *et al.* 1984). Influenza virus A has also been traced to grey seals, *Halichoerus grypus*, and harp seals, *Phoca groenlandica*, in eastern Canada (Geraci *et al.* 1984).

The second example concerns rinderpest, which remains a problem in Africa, where it has been spreading in recent years. It has been described as probably the most lethal and potentially dangerous infectious disease which affects Artiodactyla. In domestic cattle it can be controlled by vaccination and rigorous control of movements in infected areas, but these methods cannot be applied to wildlife.

Apparently there is no record of disease in hippopotamus, *Hippopotamus amphibius,* that has been attributed to rinderpest virus and so it is generally regarded as insusceptible. A survey of rinderpest neutralizing antibody was carried out on 315 hippopotamus shot in Queen Elizabeth National Park,

Uganda, in 1962 (Plowright, Laws & Rampton 1964). There are records of rinderpest outbreaks in the area in 1920/21, 1931/33, and 1944/45. No serological positives were found in animals aged 11 or less; in the age class 13–15 years 4% of the animals examined were positive; but in older animals (to 41 years) 36% were positive. This implied that the intensity of exposure in the earlier epizootics was greater than in 1944/45. The mean antibody level in hippopotamuses first affected about 30 years before was found to be at least as high as that in yearling wildebeest, infected only a few weeks or months before. Not only was this an outstanding example of the stability of such antibodies in animals, but it was possible to use the pattern of antibody levels in hippopotamus to test the criteria being used to determine age in this species (Laws 1968).

I would like to conclude by acknowledging the generous support from The Wellcome Trust to the scientific programme of the Zoological Society of London. This support, granted on a competitive basis through the normal, strict, peer review of project applications, has resulted in a mass of data and results that goes some way towards resolving the problems identified at the beginning of these introductory remarks. The novel insights into fundamental biological questions promoted by the study of exotic species have also led to several applications in the agricultural and medical sciences. This successful partnership between the Wellcome Trust and the Zoological Society, aided in addition through the Trust's help in establishing the Wellcome Laboratories of Comparative Physiology, make it entirely appropriate that this volume is published to mark the Trust's 50th Anniversary.

References

Geraci, J.R., St Aubin, D.J., Barker, I.K., Webster, R.G., Hinshaw, V.S., Bean, W.J., Ruhnke, H.L., Prescott, J.H., Early, G., Baker, A.S., Madoff, S. & Schooley, R.T. (1982). Mass mortality of harbor seals: Pneumonia associated with influenza A virus. *Science, N.Y.* **215:** 1129–31.

Geraci, J.R., St Aubin, D.J., Barker, I.K., Hinshaw, V.S., Webster, R.G. & Ruhnke, H.L. (1984). Susceptibility of grey (*Halichoerus grypus*) and harp (*Phoca groenlandica*) seals to the influenza virus and mycoplasma of epizootic pneumonia of harbor seals (*Phoca vitulina*). *Can. J. Fish. aquat. Sci.* **41:** 151–6.

Hinshaw, V.S., Bean, W.J., Webster, R.G., Rehg, J.E., Fiorelli, P., Early, G., Geraci, J.R. & St Aubin, D.J. (1984). Are seals frequently infected with avian influenza viruses? *J. Virol.* **51:** 863–5.

Laws, R.M. (1960). The southern elephant seal (*Mirounga leonina* Linn.) at South Georgia. *Norsk Hvalfangsttid.* **49:** 466–76; 520–42.

Laws, R.M. (1968). Dentition and ageing of the hippopotamus. *E. Afr. Wildl. J.* **6:** 19–52.

Laws, R.M. (1979). Monitoring whale and seal populations. *Symp. Inst. Biol.* No. 24: 115–40.

Laws, R.M., Parker, I.S.C. & Johnstone, R.C.B. (1975). *Elephants and their habitats. The ecology of elephants in north Bunyoro, Uganda.* Clarendon Press, Oxford.

Laws, R.M. & Taylor, R.J.F. (1957). A mass dying of crabeater seals, *Lobodon carcinophagus* (Gray). *Proc. zool. Soc. Lond.* **129:** 315–24.

Plowright, W., Laws, R.M. & Rampton, C.S. (1964). Serological evidence for the susceptibility of the hippopotamus (*Hippopotamus amphibius* Linnaeus) to natural infection with rinderpest virus. *J. Hyg., Camb.* **62:** 329–36.

Symp. zool. Soc. Lond. (1988) No. 60 7–28

Natural suppression of fertility

David H. ABBOTT

MRC/AFRC Comparative Physiology Group
Institute of Zoology
The Zoological Society of London
Regent's Park
London, NW1 4RY

Synopsis

The three major natural suppressors of fertility are (i) the environment, including its influence on seasonal breeding and nutritional status, (ii) lactation and (iii) social dominance. This paper will concentrate on the physiological impairments imposed by the latter, and specifically on the striking infertility found in female social subordinates. Stress is increasingly implicated in human infertility, and a greater understanding of stress-related infertility occurring naturally in mammals may prove rewarding.

Suppression of fertility as a result of social dominance is found throughout the mammalian order. High-ranking or dominant females produce more offspring than low-ranking or subordinate females. In some species, such as the marmoset monkey and the naked mole-rat, the effect is so extreme that there is only one breeding female in each group.

In subordinate female marmosets, ovulation is suppressed because of gonadotrophin insufficiency. Removal of subordinates from dominant females resulted in an increased secretion of pituitary luteinizing hormone (LH) and the onset of ovulatory cycles. When subordinates were returned to dominant females this process was reversed. The pituitary inadequacy in subordinates was due to the suppression of gonadotrophin releasing hormone (GnRH) from the hypothalamus. Infusion of exogenous GnRH by means of a minipump 'sidestepped' the hypothalamic block and stimulated pituitary LH secretion. The hypothalamic block to GnRH was not associated with high blood concentrations of cortisol or prolactin or with reduced body weight. However, it was associated with an increase in negative feedback sensitivity to oestradiol and a possible direct neural inhibition of GnRH release. Initial studies with the opiate antagonist naloxone did not implicate opioid peptides as neural mediators of GnRH suppression.

Naked mole-rats live in large underground colonies. Ovulation is suppressed in all females except the 'breeding queen'. Subordinates are the colony's workers. Removal of subordinates from dominant females resulted in the onset of their ovulatory cycles.

The naturally suppressed ovulation of subordinate female marmoset monkeys and

ZOOLOGICAL SYMPOSIUM No. 60
ISBN 0-19-854007-8

naked mole-rats may yield valuable information about the physiological translation of stress into infertility.

Introduction

Socially induced infertility, or social contraception (Abbott 1984), is now recognized in free-living populations of many animal species, e.g., the gelada baboon (Dunbar 1980), yellow baboon (Wasser & Barash 1983), tamarin monkey (Garber, Moya & Malaga 1984; Terborgh & Goldizen 1985), marmoset monkey (Hubrecht 1984; Stevenson & Rylands in press), silver-backed jackal (Moehlman 1983), dwarf mongoose (Rood 1980), prairie vole (Carter, Getz & Cohen-Parsons 1986), and naked mole-rat (Jarvis 1984; Brett 1986). This natural infertility syndrome occurs when high-ranking or dominant individuals suppress reproduction in low-ranking or subordinate individuals (Harcourt 1987) or when parents suppress reproduction in their maturing offspring (see Kleiman 1977).

Socially induced reproductive suppression occurs in its most extreme form when a female accepts help in rearing her offspring to maximize their chances of survival (Wasser & Barash 1983). Typical examples are found in several species of marmoset and tamarin monkeys; one female breeds in each social group and all the remaining adults and older offspring help carry and care for her young (Epple 1975; Abbott & Hearn 1978; French, Abbott & Snowdon 1984; Garber *et al.* 1984; Terborgh & Goldizen 1985). When such extreme social suppression of reproduction exists, the reproduction of potential female competitors is suppressed by (i) inhibited ovulation, which occurs as a result of the stress incurred from the continual behavioural subordination (as in the marmoset monkey; Abbott 1984), (ii) interruption of social and sexual behaviour by the dominant female (as in the wolf; Mech 1970; Seal, Plotka, Packard & Mech 1979), or (iii) direct or indirect intervention by the dominant female contributing to the early demise of their offspring (as in the wild dog; Malcolm & Marten 1982). However, social contraception is not the only form of natural fertility suppression.

There are two other major types of naturally occurring infertility in free-living mammals, excluding infertility due to immaturity, old age, disease or congenital abnormalities. The first type is produced by environmental causes. These consist of (i) the changing seasons of the year, with the consequent variations in climate, daylength and rainfall, and (ii) food availability. In temperate regions of the world, where a hospitable climate and readily available food occur only at certain times of the year, many mammals have adopted a seasonal breeding strategy. In excellent examples such as the red deer (Lincoln & Short 1980) and the sheep (Karsch, Bittman, Foster, Goodman, Legan & Robinson 1984), breeding capability is suppressed during the spring and summer months and maximized during autumn and winter to give a synchronized birthing period in spring and early summer, when the grass begins to grow and the survival rate of the offspring is optimal. This ability to suppress reproduction during a particular time of the year has been called 'nature's contraceptive' by Lincoln & Short (1980), who among many have shown it to be brought about physiologically by changes in daylength through the year, causing alterations in the secretion pattern of melatonin from the pineal gland (for a review see Bittman 1984). Melatonin does not appear to affect directly the reproductive 'shut down' (Hastings & Herbert 1986), but acts through an as yet unknown neural mechanism, which may be associated with endogenous opioid peptides in the brain (Hastings, Herbert, Martensz & Roberts 1985; Roberts, Martensz, Hastings & Herbert 1985) and, as in birds, with secretions from the thyroid gland (Follett & Nicholls 1984; Nicholls, Goldsmith & Dawson 1984). In physiological terms, the daylength-induced changes in melatonin secretion cause reduced secretion of gonadotrophin releasing hormone (GnRH) from the hypothalamus in the brain, which results in insufficient stimulation of synthesis and release of gonadotrophins—luteinizing hormone (LH) and follicle stimulating hormone (FSH)—from the anterior pituitary gland. This in turn leads to insufficient stimulation of the gonads and finally to gonadal failure, such as suppressed ovulation or reduced sperm production.

A shortage of good quality food can reduce fertility in free-living mammals, e.g., vervet monkeys (Lee 1987). The physiological mechanisms invoked by inadequate nutrition are not fully understood, but as in seasonally breeding mammals, GnRH secretion is inhibited (Dubey, Cameron, Steiner & Plant 1986).

The second major type of naturally occurring infertility is found in females during lactation (Short 1976). In women and many other female mammals with a relatively long life span, a sucking infant suppresses ovulation (Short 1976; Howie & McNeilly 1982; McNeilly in press a). Once the suppressive effects of high concentrations of steroid hormones in the blood disappear after birth (McNeilly 1979, in press b), a suckling rate of more than five times a day ($\geqslant$ 10 min per feed) will maintain the inhibition of gonadotrophin secretion from the pituitary gland, apparently by suppressing the release of GnRH from the hypothalamus (Glasier, McNeilly & Baird 1986). The infertile period ends only when suckling has decreased to the point at which GnRH is no longer inhibited. However, in women, a return to ovulation can be hastened by introducing food supplements to the baby's diet. Howie & McNeilly (1982) found that the duration and frequency of suckling dropped dramatically when food supplements were introduced, and over half of the 27 women studied had resumed ovulating within 16 weeks.

Because prolactin concentrations in the blood were greatly elevated during lactation, it was long thought that prolactin itself inhibited GnRH release (McNeilly 1984). However, any such role now appears to be minimal (Schallenberger, Richardson & Knobil 1981) and, certainly in women, the blood concentrations of prolactin during lactation are simply related to suckling activity (McNeilly, Glasier & Howie 1985; McNeilly in press b). In pigs and rats, suckling may suppress GnRH output by increasing hypothalamic opiate activity (Sirinathsinghji & Martini 1984; Mattioli, Conte, Graleati & Seren 1986), but there is as yet no direct evidence for a similar effect in women.

Short (1976) proposed that during man's primitive past the number of offspring produced by a female in her lifetime was limited by lactational amenorrhoea, which thus effectively controlled the population size. In the nomadic society of the !Kung hunter gatherers of southern Africa, lactational infertility can last for four years (Short 1976; Konner & Worthman 1980). Study of this population has also illustrated the dramatic effects of supplementary feeding in shortening the contraceptive effect of breast feeding (Short 1976).

This review of the environmental and lactational causes of natural infertility has been kept brief because the main topic of the paper is the mechanisms that mediate infertility induced by social dominance. Attention will be focussed on females, in which the suppressive effects on fertility are most striking (Abbott 1984). The study of natural stress-related infertility is particularly important today because stress is becoming increasingly linked with GnRH-related infertility in women (Dunbar 1985). The defined and repeated amounts of behavioural stress imposed by dominant animals on their subordinates in free-living or laboratory-housed populations may well provide some insight into the mechanisms by means of which stress suppresses GnRH secretion and infertility.

Consequently, this paper will focus on two mammalian species in which, in each social group, only one female ovulates and reproduces and the number of breeding males is restricted to one or two. The first is a primate, the common marmoset monkey (*Callithrix jacchus*). The second is a rodent, the eusocial naked mole-rat (*Heterocephalus glaber*). The following discussion will deal with the endocrinological and neuro-endocrinological mechanisms that produce infertility in social subordinates in these species.

Socially induced infertility in subordinate female marmoset monkeys

Subordinate females in social groups

To study infertility in subordinate female marmoset monkeys, the composition of laboratory social groups was standardized. Unrelated post-pubertal offspring, born in captivity, were removed from their families and placed together in groups, usually consisting of three males and three females

(Abbott & Hearn 1978; Abbott 1986). The social dominance hierarchy of each group became established within 2–3 days, resulting in the pair-bonding of one dominant (Rank 1) male and one dominant (Rank 1) female (Abbott & Hearn 1978; Abbott 1984). Subordinate females (Ranks 2 and below) were identified from the submissive way in which they behaved towards the dominant female, regardless of any aggression shown by the latter in initiating the interaction or responding to it (Abbott & Hearn 1978).

Repeated checks on social status were made 2–5 and 6–9 months after group formation (Abbott 1986). These confirmed the ranks to which the females were initially assigned. The findings described in this paper were based on 12 social groups which had been formed for up to two years.

The management of marmoset monkeys in captivity, and the reproductive endocrinology of females, have already been well described (Hearn 1983). To maintain the static composition of the social groups, dominant females were given an injection of synthetic prostaglandin-F2α (Summers, Wennink & Hodges 1985) to terminate pregnancy, usually 22–28 days after conception. Routine blood samples were taken from all females twice weekly (Abbott 1984).

Failures in the reproductive endocrinology of subordinate females

Ovarian cycles

Dominant females showed normal ovulatory cycles, but 18 of the 22 subordinate females showed no such activity, their plasma progesterone concentrations never rising to the values ($\geq$ 10 ng/ml) that would have indicated ovulation (Harding, Hulme, Lunn, Henderson & Aitken 1982). However, four of the subordinate females, in four of the 12 groups, showed evidence of incomplete ovulatory cycles. Each demonstrated a rise in plasma progesterone ($>$ 10 ng/ml) for 1–10 days, but in none was it maintained for the 19–20 days of the normal luteal phase (Harlow, Hearn & Hodges 1984). One, and only one, of these four subordinate females had a similarly incomplete cycle nine months later (Abbott 1986).

Only acyclic subordinate females are considered in the following sections.

Pituitary gonadotrophin secretion

Suppression of ovarian cycles in subordinate females immediately suggested an examination of circulating gonadotrophin concentrations to reveal any reduced pituitary output. Because there is no FSH assay for the marmoset, only bioactive plasma LH concentrations were determined (Harlow *et al.* 1984). Basal plasma LH concentrations in subordinate females (4.1 $\pm$ 0.5 mIU/ml, 85 samples from five animals) were significantly ($p < 0.05$) lower than in dominant females (16.8 $\pm$ 1.1 mIU/ml, 85 samples from five

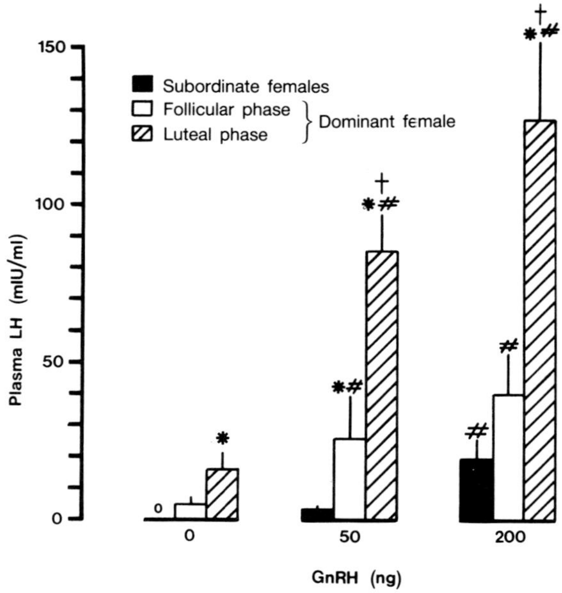

Fig. 1. Maximum plasma LH elevations (mean ± s.e.m.) induced by 0, 50 and 200 ng GnRH in subordinate and dominant female marmosets.
○ = LH values below assay sensitivity (< 2 mIU/ml);
✱ = $p < 0.05$ versus subordinate;
† = $p < 0.05$ versus follicular phase, dominant;
# = $p < 0.05$ versus 0 ng (Duncan's Multiple Means test following ANOVA).

animals; all non-pregnant and in the luteal phase). Additional evidence of insufficient LH output from the pituitary of subordinate females came from a series of GnRH challenges (Abbott, Hodges & George in press; Fig. 1). The maximum LH responses of subordinate females to GnRH (50 ng) were substantially less than those of dominant females in either the follicular or luteal phase of the ovarian cycle; and at a GnRH dose of 200 ng they were less than those of dominant females in the luteal phase. The highest dose of GnRH used (200 ng) produced the largest LH rise from all females. However, luteal phase LH values in dominant females, under GnRH stimulation, were significantly higher than those of animals in their follicular phase or in subordinate females (Fig. 1). In women, also, LH responsiveness to a GnRH challenge is higher in the luteal phase than in the follicular phase (Yen 1978). Basal (0 ng GnRH) luteal-phase LH values were significantly greater than those in subordinates, but not than those associated with the follicular phase. Luteal phase concentrations of progesterone in marmosets—in contrast to man—may not exert strong suppression on gonadotrophin secretion.

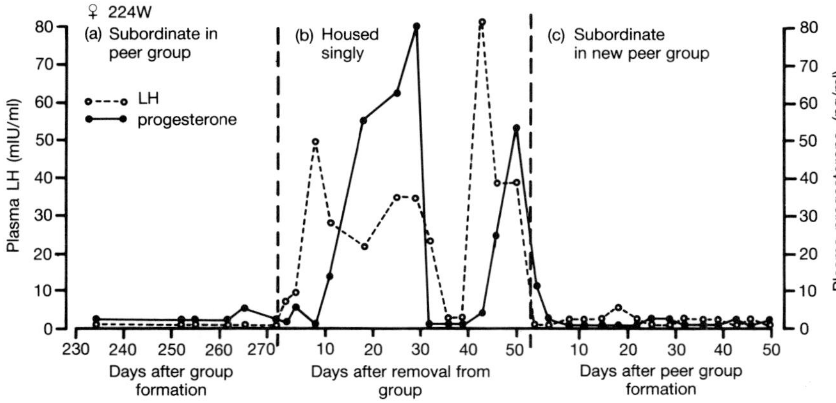

Fig. 2. Plasma progesterone and LH concentrations in female marmoset 224W, (a) while subordinate in a social group, (b) housed singly and (c) after becoming subordinate in a new social group.

Would removal of subordinates from the inhibiting influence of their dominant females result in a rise of pituitary LH secretion and the onset of ovulation?

Suppression of LH secretion and ovulation is dependent on subordinate social status

When four long-term subordinate females were removed from their social groups and housed singly, their plasma LH concentrations rose above 2 mIU/ml (the sensitivity limit of the LH bioassay) within 1–4 days ($n = 3$) and 7 days ($n = 1$). A pre-ovulatory LH surge in all four females was followed by ovulation 9–46 days after removal from their groups (see Fig. 2 for an example). Normal ovarian cycles continued until 38–77 days, when the females were placed in new social groups and reverted to subordinate social status. Within 1–4 days of returning to this status, all four females showed plasma LH levels that had fallen to close to or below 2 mIU/ml (e.g., Fig. 2). No further pre-ovulatory LH surges or ovulations took place. Females in the luteal phase of their ovarian cycle ($n = 2$) or at the peri-ovulatory period ($n = 1$) when subordinate social status was re-imposed suffered from premature termination of their cycle, with plasma progesterone concentrations falling below 10 ng/ml within 5–10 days. Apparently, as in female rhesus monkeys (Hutchison & Zeleznik 1984), the corpus luteum in female marmoset monkeys requires gonadotrophic support to function normally.

This experiment clearly demonstrated that suppression of LH secretion

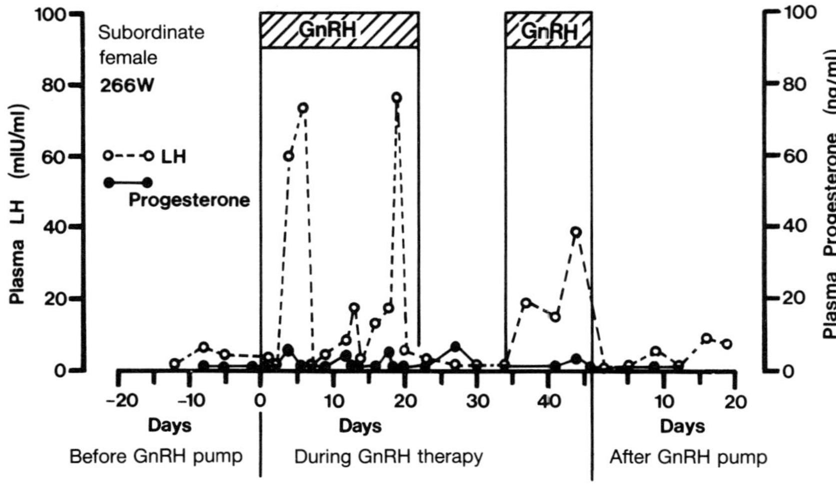

Fig. 3. Plasma LH concentrations in subordinate female marmoset 266W before, during and after GnRH treatment (1 μg GnRH every 2.4 h, subcutaneously).

and ovulation was entirely due to subordinate social status. Removal of subordinate status resulted in a rapid 'switch on' of the reproduction system. Imposition of subordinate status resulted in an equally rapid 'switch off'. If this 'switch on' and 'switch off' merely reflected what was happening to GnRH release from the hypothalamus, could GnRH treatment of subordinate females remaining in their social groups stimulate LH secretion and induce ovulation?

Infusion of GnRH by minipump

Ten long-term (> 2 months) subordinate females were fitted with a lightweight backpack and minipump assembly connected to a subcutaneously implanted cannula (Ruiz de Elvira & Abbott 1986). GnRH was infused in a pulsatile manner, 1 μg approximately every 2 h, for periods ranging from 6 to 22 days. As illustrated in Fig. 3, GnRH caused an increase in the plasma LH concentrations. In this example, two LH peaks (73 and 76 mIU/ml) were induced by GnRH after 6 and 19 days respectively. Both peaks were within the normal pre-ovulatory surge values of 60–340 mIU/ml (Harlow *et al.* 1984). Before and after both periods of GnRH treatment and during the gap between them, plasma LH concentrations remained low (Fig. 3).

GnRH treatment of subordinate females remaining in their social groups apparently succeeded in stimulating pituitary LH secretion (thus implicating suppressed GnRH secretion as the cause of gonadotrophin insufficiency), but ovulation was not induced. The plasma progesterone concentrations all remained below 10 ng/ml (Fig. 3). Longer periods of GnRH infusion or alterations in the GnRH dose and the interval between GnRH pulses may be required before ovulation is obtained.

What is the cause of suppressed GnRH release in subordinate females?

Suppression of GnRH: cortisol, prolactin and low body weight

Cortisol and prolactin concentrations in the blood are generally elevated in times of stress. Such elevations have been implicated in the suppression of LH secretion in primates, presumably via suppressed GnRH (Keverne 1979; Dubey & Plant 1984). However, plasma cortisol and prolactin concentrations are not elevated in subordinate female marmosets (Abbott, McNeilly, Lunn, Hulme & Burden 1981).

Low body weight in subordinate females can also be ruled out as a contributory cause of GnRH suppression because dominant and subordinate female body weights do not differ significantly (Abbott 1986).

Suppression of GnRH: increased sensitivity to oestradiol negative feedback

Ovarian steroid hormones, such as oestradiol, exert a negative feedback control on GnRH and LH secretion (Everitt & Keverne 1986). Removal of this ovarian control results in a striking and sustained increase in GnRH and LH secretion. To test whether an increased sensitivity to the negative feedback effects of hormones secreted from the quiescent ovaries of subordinate females played a role in the social suppression of LH secretion, and hence probably of GnRH secretion, five subordinate and three dominant females were ovariectomized and their LH responses measured. Plasma LH concentrations in the dominant females rose to a post-ovariectomy maximum of 71.4 ± 8.7 mIU/ml within one week (Fig. 4). High LH levels were maintained thereafter. In contrast, plasma LH concentrations in subordinate females took longer to rise and, even three weeks after ovariectomy, they had not reached the LH values attained by dominant females within one week (Fig. 4). After one year, this difference in LH between dominant and subordinate ovariectomized females was still evident (63.6 ± 4.0 mIU/ml and 14.2 ± 1.7 mIU/ml, respectively).

A low-dose oestradiol implant given to dominant and subordinate ovariectomized females induced circulating oestradiol levels of approximately 350 pg/ml (Abbott, George, Hodges & Cottingham in prep.). This was far less than the concentration (1 ng/ml) produced by Hodges (1978) to suppress plasma LH in ovariectomized female marmosets. The low oestradiol dose was intended to be too low to suppress LH secretion in ovariectomized dominant females. This intention succeeded, the plasma LH concentrations being 50.0 ± 7.2 and 71.9 ± 10.8 mIU/ml immediately before and 3 weeks after implantation, respectively. In ovariectomized females, however, the plasma LH values were 30.8 ± 8.9 and 3.1 ± 1.1 mIU/ml immediately before and one week after implantation, respectively. Thus subordinate

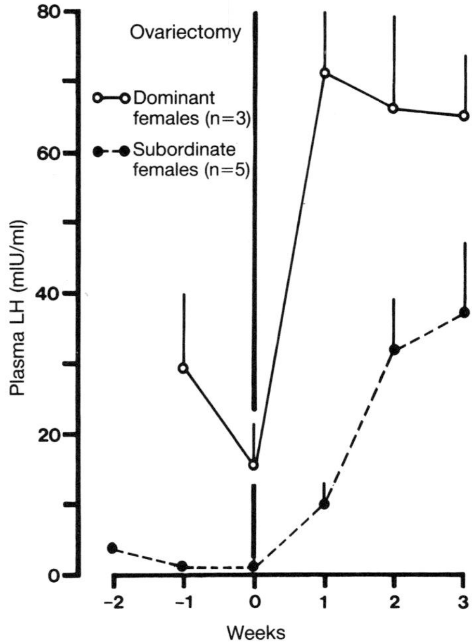

Fig. 4. Plasma LH concentrations (mean ± s.e.m.) after ovariectomy in three dominant and five subordinate female marmosets.

females were more sensitive than dominant females to the negative feedback effects of oestradiol on LH secretion. Could this be the mechanism suppressing LH levels in subordinate females? It certainly might explain the poor LH response of subordinates to ovariectomy and their rapid 'switch off' of LH secretion in response to a low-dose of oestradiol, but it did not convincingly explain why LH levels in subordinate females remained low after ovariectomy.

Could there be a neural component, independent of oestradiol negative feedback, to the mechanism suppressing GnRH and LH secretion?

Suppression of LH secretion in ovariectomized females is dependent on subordinate social status

To maintain constant oestradiol levels in the blood, five ovariectomized females were isolated from their social group for seven weeks, and were then returned to subordinate states, either ($n = 4$) within their own group or ($n = 1$) in a new group. In four of the five isolated females, the plasma LH rose to the concentrations found in dominant ovariectomized females within 30 days (see the example in Fig. 5). Within 30 days of returning these ovariectomized females to their original subordinate social status, the

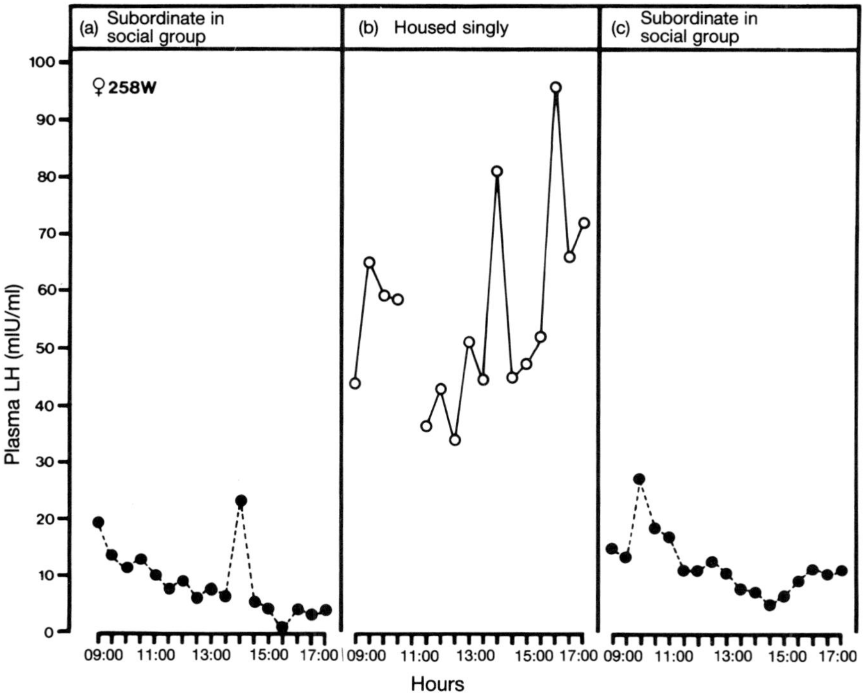

Fig. 5. Plasma LH concentrations in ovariectomized female marmoset 258W: (a) when subordinate in a social group (● --- ●); (b) while housed singly for 30 days (○ — ○), and (c) after returning to subordinate status in her original social group for 30 days (● -- ●). Blood samples were taken every 30 min between 09:00 and 17:00 h on the days concerned.

plasma LH had returned to its previously low concentration (Fig. 5). A change in negative feedback sensitivity to circulating steroids from extra-gonadal sources (e.g. the adrenals) may completely explain these LH results; on the other hand this experiment raises the possibility that a steroid-independent neural mechanism may play an important role in mediating GnRH and LH suppression imposed through subordinate social status.

Neuroendocrine suppression of GnRH in subordinate females: role of opioids

The endogenous opioid peptides have been implicated directly (Grossman, Moult, Gaillard, Delitalia, Toff, Rees & Besser 1981) and indirectly (Bicknell 1985) in the suppression of hypothalamic GnRH secretion. The use of the opioid receptor blocker, naloxone, to stimulate LH secretion in women with hypothalamic amenorrhoea has implicated the endogenous opioid

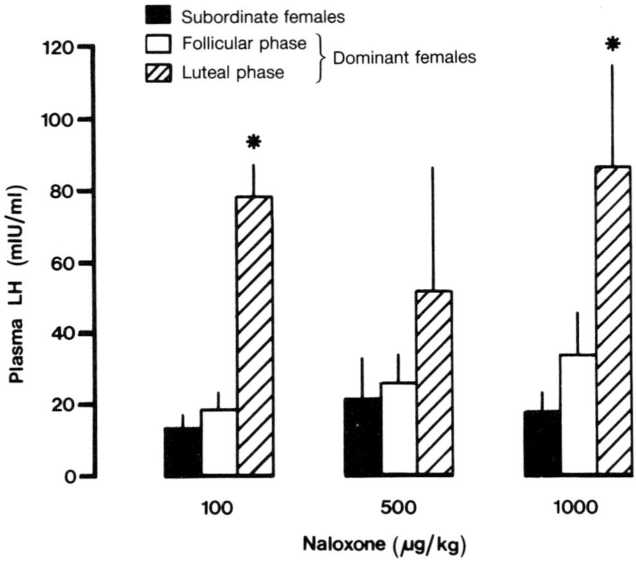

Fig. 6. Maximum plasma LH elevations (mean ± s.e.m.) induced by 100, 500 and 1000 μg naloxone (Narcan, Endo Laboratories Inc.) in subordinate and dominant female marmosets. * = $p < 0.05$ versus subordinate (Duncan's Multiple Means test following ANOVA).

peptides in GnRH-related infertility (Quigley, Sheehan, Casper & Yen 1980; Yen 1986). Recently, increased concentrations of β-endorphin in the cerebrospinal fluid have been linked with subordinate social status in male talapoin monkeys (Martensz, Vellucci, Herbert & Keverne 1986). Subordinate and dominant female marmosets were therefore given naloxone to determine whether suppressed LH secretion in subordinates was related to opioid inhibition.

Dominant females in the luteal phase of their ovarian cycle differed significantly from subordinates in showing a greater LH response to naloxone (Fig. 6). It appears that opioid peptides assist in controlling GnRH secretion in the luteal phase of the ovarian cycle in the marmoset, in rhesus monkeys (Van Vugt, Lam & Ferin 1984) and in women (Ropert, Quigley & Yen 1981). They do not, however, have the same suppressive effect in the follicular phase—in rhesus monkeys (Van Vugt, Bakst, Dyrenfurth & Ferin 1983) and women (Quigley & Yen 1980)—or in acyclic subordinates. Short-term blockade of opiate receptors may be insufficient to permit a substantial release of GnRH in subordinate females. On the other hand, endogenous opioids may not play a significant role in modulating GnRH output in the absence of gonadal hormones (Kesner, Kaufman, Wilson, Kuroda & Knobil 1986). Further work is required to identify neuroendocrine mediators of GnRH suppression in subordinate female marmosets.

Socially induced infertility in subordinate female naked mole-rats

Subordinate females in social groups

The eusocial naked mole-rat has taken social contraception one stage further than the marmoset and has developed an extreme division of labour between dominant breeding animals and subordinate non-breeding animals (Jarvis, 1981, in press; Brett 1986; Lacey & Sherman in press). In each large underground colony, in a way analogous to that of the social insects (Jarvis 1984), subordinates maintain, defend and expand the tunnel system and provision the colony's nesting chamber with food. The larger subordinates have been associated mainly with defence, and the smaller subordinates mainly with tunnel maintenance and food provisioning (Jarvis 1981; Lacey & Sherman in press).

In the laboratory, colonies are housed in a series of perspex or glass tunnels and chambers (see for example Brett 1985). Endocrine monitoring of individual females is based on urine samples collected from each colony's communal toilet chamber. The preliminary hormonal data presented in this paper come from four captive colonies established in the laboratory for 1–3 years.

Failures in the reproductive endocrinology of subordinate females

Reproductive development

Only one female, the breeding 'queen', reproduces in each colony of naked mole-rats (Jarvis 1981, 1984, in press; Lacey & Sherman in press). The 'queen' is also the only female to show sexual behaviour with the one or two breeding males in each colony (Jarvis 1981, 1984; Lacey & Sherman in press). The reproductive development of subordinate females appears to be arrested at the pre-pubertal stage because their vaginas remain imperforate (Jarvis 1981, 1984; Liddell, Abbott, Faulkes, George & Jarvis in press), their external genitalia are poorly differentiated, their ovaries are small and show little follicular development (Kayanja & Jarvis 1971), and the animals do not attain full body size (Jarvis 1981, 1984).

Ovarian cycles

Only the breeding 'queen' ovulated and became pregnant in a colony monitored for endocrine function. Urinary progesterone concentrations were negligible in all non-breeding females, in contrast to the higher progesterone values invariably found in the breeding 'queen' (Fig. 7).

This is the first time that suppression of ovulation has been confirmed, hormonally, in subordinate naked mole-rats. Could removal of subordinate social status terminate suppression as in the marmoset?

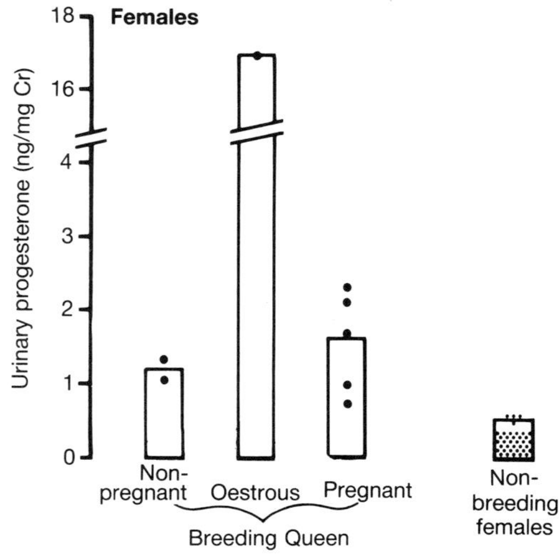

Fig. 7. Urinary progesterone concentrations in females from one captive colony of naked mole-rats. Each point represents a single sample from an individual. Individual females were sampled between one and eight times. The histogram represents the mean for each category.

Suppression of reproduction is dependent on subordinate social status
Removal of subordinate social status was affected on three occasions. In the first instance, the original colony imported from Kenya was divided and 18 animals were separated from the breeding animals to form a new colony. The first female to hold the dominant position in the new colony was the deposed breeding female from the original colony. She had not held breeding status for almost two years. She was challenged by another previously subordinate female after 1–2 weeks. Both died of their wounds, but perforate vaginas had already developed. A third previously subordinate female took over the dominant position without any further fighting and became pregnant three months after the new colony was separated from the original one. She produced live pups 10 weeks later and rapidly attained full body size.

On the two other occasions, two male–female pairs of subordinate naked mole-rats were removed from their colonies and were established as independent pairs. Both females became pregnant (after one and five months) and have since repeatedly produced live pups at birth. The return of reproductively active females to their original colonies has not been attempted because of the extremely aggressive response of naked mole-rats to animals from colonies other than their own (Lacey & Sherman in press), and because of the lethal aggression displayed by reproductively active females towards each other. Nevertheless, the evidence is convincing that reproductive suppression in female naked mole-rats is due to subordinate social status.

Discussion

The social suppression of fertility in subordinate female marmoset monkeys and naked mole-rats provides two powerful examples of how social status can determine reproductive function. The extreme development of this trait in these two species, rendering all subordinates infertile, permitted an examination of the physiological mechanisms that translate behavioural subordination into infertility.

Such social subordinates appear to have an endocrine dysfunction similar to that accompanying 'functional hypothalamic amenorrhoea of the psychogenic type' in women (Yen 1986). Reduced GnRH output is the basic hypothalamic defect in stress-related infertility of this kind in women (Leyendecker & Wildt 1983; Crowley, Filicori, Spratt & Santoro 1985) and may be brought about through elevations of hypothalamic opioid or dopamine activity (Quigley *et al.* 1980; Grossman *et al.* 1981; Yen 1986).

Because they do not reach puberty subordinate female naked mole-rats seem influenced by a combination of functional hypothalamic amenorrhoea and primary hypogonadotrophic hypogonadism. On the other hand, subordinate female marmosets seem influenced by a combination of functional hypothalamic amenorrhoea and secondary hypogonadotrophic hypogonadism, because females can pass puberty and undergo ovarian cycles before becoming infertile subordinates in an adult social group (Abbott 1984). The difference in intensity between the social suppression of fertility in the two species may reflect a need for the tight control of fertility in colonies of naked mole-rats, whose subordinates provide the work and defence force (Brett 1986; Lacey & Sherman in press) and do not migrate to other colonies (Brett 1986). The divisions of labour in subordinate marmosets are, in contrast, much less striking, and migration to other groups is common (Terborgh & Goldizen 1985).

Increased hypothalamic suppression of GnRH secretion by endogenous opioid peptides has been implicated as the neuroendocrine mediator of all three major sources of natural infertility—environment, lactation, and social dominance—respectively by Roberts *et al.* (1985) in the hamster, Sirinathsinghji & Martini (1984) in the rat, and Martensz *et al.* (1986) in the talapoin monkey. However, in all but lactational infertility, blockade of opioid receptors with naloxone has not yet provided evidence of a functional connection between elevated concentrations of endogenous opioid peptides in the brain and GnRH suppression (see also Fig. 6).

An increased ability of oestradiol to suppress gonadotrophin secretion has also been implicated in the physiological mechanism of natural reproductive suppression, as illustrated in the seasonally anoestrous ewe (Legan & Karsch 1980), the suckling human female (Baird, McNeilly, Sawers & Sharpe 1979) and socially subordinate marmoset monkeys (see above and Fig. 4). It has been argued that the decreased activity of the hypothalamic–pituitary axis, evident after ovariectomy in the anoestrous ewe (Goodman, Bittman, Foster & Karsch 1982) and subordinate marmoset (Figs 4 and 5), makes this system more sensitive to any inhibiting input (Lincoln & Short 1980). However, evidence is accumulating that in seasonally suppressed breeders such as the sheep (Robinson, Radford & Karsch 1985; Meyer & Goodman 1986) and socially suppressed breeders such as the marmoset (Fig. 5), steroid-independent neural mechanisms may provide the underlying control of GnRH secretion from the hypothalamus, in addition to and separate from the steroid-dependent effect manifested as an increased ability of oestradiol to suppress gonadotrophin secretion. It is interesting that seasonal breeding, social suppression of fertility, and possibly lactational infertility may all share common neuroendocrine mechanisms.

Pheromones from dominant females may initiate the neuroendocrine suppression of reproduction in subordinate female marmoset and tamarin monkeys (Epple & Katz 1984) and naked mole-rats (Jarvis 1981, 1984). Certainly in male lesser mouse lemurs, air circulated through the urine from dominant males is sufficient to impair reproduction in subordinate males (Schilling, Perret & Predine 1984). In female mice, exposure to urine from mature females delays puberty (Cowley & Wise 1972; Massey & Vandenbergh 1980) and induces anoestrus or the 'Whitten effect' (Whitten 1959). Pheromones in the urine of adult female prairie voles similarly suppress puberty in female offspring (Carter *et al.* 1986). Further study of pheromonal action in the social suppression of reproduction may help elucidate the neural mechanisms that influence the hypothalamus and GnRH secretion.

Delay of sexual maturation and suppression of ovulation are not the only manifestations of socially induced female infertility (Wasser & Barash 1983). Other reproductive defects, such as failure of implantation or spontaneous abortion (e.g. in the mouse: Lloyd & Christian 1969) may also occur, but reliable detection in many species is difficult. These post-ovulatory failures may be related to disrupted gonadotrophin secretion or to other factors that reduce luteotrophic support, such as inhibited prolactin secretion in the mouse (Keverne 1983). Further investigation of socially induced infertility may well serve to increase our understanding of the mechanisms that control mammalian reproduction.

Acknowledgements

The author would like to thank his colleagues, including J. K. Hodges, L. M. George, M.-C. Ruiz de Elvira, C. E. Liddell, S. F. Ferrari, J. W. Sheffield,

A. J. Leigh and B. R. Ferreira for their collaboration in some of the work reported here; M. J. Llovet, T. D. Noble and the animal technical staff for care and maintenance of the animals; and T. J. Dennett and M. J. Walton for preparation of the figures. This work was supported by grants for the Association for the Study of Animal Behaviour, the Nuffield Foundation and the University of London and an MRC/AFRC programme grant.

References

Abbott, D.H. (1984). Behavioural and physiological suppression of fertility in subordinate marmoset monkeys. *Am. J. Primatol.* **6:** 169–86.

Abbott, D.H. (1986) Social suppression of reproduction in subordinate marmoset monkeys (*Callithrix jacchus jacchus*). In *A primatologia no Brasil* **2:** 1–16. (Ed. De Mello, M.T.). Sociedade Brasileira de Primatologia, Brasilia.

Abbott, D.H., George, L.M., Hodges, J.K. & Cottingham, P. (In preparation). *Reduced LH response to ovariectomy and increased sensitivity to oestradiol negative feedback in subordinate female marmoset monkeys.*

Abbott, D.H. & Hearn, J.P. (1978). Physical, hormonal and behavioural aspects of sexual development in the marmoset monkey, *Callithrix jacchus. J. Reprod. Fert.* **53:** 155–66.

Abbott, D.H., Hodges, J.K. & George, L.M. (In press). Social status controls LH secretion and ovulation in female marmoset monkeys (*Callithrix jacchus*) *J. Endocr.*

Abbott, D.H., McNeilly, A.S., Lunn, S.F., Hulme, M.J. & Burden, F.J. (1981). Inhibition of ovarian function in subordinate female marmoset monkeys (*Callithrix jacchus jacchus*). *J. Reprod. Fert.* **63:** 335–45.

Baird, D.T., McNeilly, A.S., Sawers, R.S. & Sharpe, R.M. (1979). Failure of oestrogen induced discharge of luteinizing hormone in lactating women. *J. clin. Endocrin. Metab.* **49:** 500–9.

Bicknell, R.J. (1985). Endogenous opioid peptides and hypothalamic neuroendocrine neurones. *J. Endocr.* **107:** 437–46.

Bittman, E.L. (1984). Melatonin and photoperiodic time measurement: evidence from rodents and ruminants. In *The pineal gland*: 155–92. (Ed. Reiter, R.J.). Raven Press, New York.

Brett, R.A. (1985). Captive breeding and management of naked mole rats (*Heterocephalus glaber*). *Proc. Symp. Ass. Br. wild Anim. Keepers* **10:** 6–11.

Brett, R.A. (1986). *The ecology and behaviour of the naked mole rat* (Heterocephalus glaber). Ph.D. Thesis: University of London.

Carter, C.S., Getz, L.L. & Cohen-Parsons, M. (1986). Relationships between social organisation and behavioural endocrinology in a monogamous mammal. *Adv. Study Behav.* **16:** 109–45.

Cowley, J.J. & Wise, D.R. (1972). Some effects of mouse urine on neonatal growth and reproduction. *Anim. Behav.* **20:** 499–506.

Crowley, W.F. Jr., Filicori, M., Spratt, D.I. & Santoro, N.F. (1985). The physiology of gonadotrophin-releasing hormone (GnRH) secretion in men and women. *Rec. Prog. Horm. Res.* **41:** 473–526.

Dubey, A.K., Cameron, J.L., Steiner, R.A. & Plant, T.M. (1986). Inhibition of gonadotrophin secretion in castrated male rhesus monkeys (*Macaca mulatta*) induced by dietary restriction: analogy with the prepubertal hiatus of gonadotrophin release. *Endocrinology* **118**: 518–25.

Dubey, A.K. & Plant, T.M. (1984). Sustained hypercortisolaemia interrupts open-loop gonadotrophin secretion in male rhesus monkeys. *Biol. Reprod.* **30** (Suppl. 1): 72.

Dunbar, R.I.M. (1980). Determinants and evolutionary consequences of dominance among female Gelada baboons *Behav. Ecol. Sociobiol.* **7**: 253–65.

Dunbar, R.I.M. (1985). Stress is a good contraceptive. *New Scient.* No. 1439: 16–8.

Epple, G. (1975). The behaviour of marmoset monkeys (Callithricidae). In *Primate Behavior. Developments in field and laboratory research* **4**: 195–239. (Ed. Rosenblum, L.A.). Academic Press, New York, San Francisco, London.

Epple, G. & Katz, Y. (1984). Social influences on oestrogen excretion and ovarian cyclicity in saddle back tamarins (*Saguinus fuscicollis*). *Am. J. Primatol.* **6**: 215–27.

Everitt, B.J. & Keverne, E.B. (1986). Reproduction. In *Neuroendocrinology*: 472–537. (Eds Lightman, S.L. & Everitt, B.J.). Blackwell Scientific Publications, Oxford.

Follett, B.K. & Nicholls, T.J. (1984). Photorefractoriness in Japanese quail: possible involvement of the thyroid gland. *J. exp. Zool.* **232**: 573–80.

French, J.A., Abbott, D.H. & Snowdon, C.S. (1984). The effect of social environment on oestrogen excretion, scent marking and sociosexual behaviour in tamarins (*Saguinus oedipus*). *Am. J. Primatol.* **6**: 155–67.

Garber, P.A., Moya, L. & Malaga, C. (1984). A preliminary field study of the moustached tamarin monkey (*Saguinus mystax*) in north eastern Peru: questions concerned with the evolution of a communal breeding system. *Folia primatol.* **42**: 17–32.

Glasier, A., McNeilly, A.S. & Baird, D.T. (1986). Induction of ovarian activity by pulsatile infusion of LHRH in women with lactational amenorrhoea. *Clin. Endocr.* **24**: 243–52.

Goodman, R.L., Bittman, E.L., Foster, D.L. & Karsch, F.J. (1982). Alterations in the control of luteinizing hormone pulse frequency underlie the seasonal variation in oestradiol negative feedback in the ewe. *Biol. Reprod.* **27**: 580–9.

Grossman, A., Moult, P.J.A., Gaillard, R.C., Delitalia, G., Toff, W.D., Rees, L.H. & Besser, G.M. (1981). The opioid control of LH and FSH release: effects of a met-enkephalin analogue and naloxone. *Clin. Endocr.* **14**: 41–7.

Harcourt, A.H. (1987). Dominance and fertility among female primates. *J. Zool., Lond.* **213**: 471–87.

Harding, R.D., Hulme, M.J., Lunn, S.F., Henderson, C. & Aitken, R.J. (1982). Plasma progesterone levels throughout the ovarian cycle of the common marmoset (*Callithrix jacchus*). *J. med. Primatol.* **11**: 43–51.

Harlow, C.R., Hearn, J.P. & Hodges, J.K. (1984). Ovulation in the marmoset monkey: endocrinology, prediction and detection. *J. Endocr.* **103**: 17–24.

Hastings, M.H. & Herbert, J. (1986). Endocrine rhythms. In *Neuroendocrinology*: 49–102. (Eds Lightman, S.L. & Everitt, B.J.). Blackwell Scientific Publications, Oxford.

Hastings, M.H., Herbert, J., Martensz, N.D. & Roberts, A.C. (1985). Melatonin and the brain. *Ciba Found. Symp.* (N.S.) **117:** 57–77.

Hearn, J.P. (1983). The common marmoset (*Callithrix jacchus*). In *Reproduction in New World Primates*: 181–215. (Ed. Hearn, J.P.). MTP Press, Lancaster.

Hodges, J.K. (1978). Effects of gonadectomy and oestradiol treatment on plasma luteinizing hormone concentrations in the marmoset monkey, *Callithrix jacchus. J. Endocr.* **76:** 271–81.

Howie, P.W. & McNeilly, A.S. (1982). Effect of breast-feeding patterns on human birth intervals. *J. Reprod. Fert.* **65:** 545–57.

Hubrecht, R.C. (1984). Field observations on group size and composition of the common marmoset (*Callithrix jacchus jacchus*) at Tapacura, Brazil. *Primates* **25:** 13–21.

Hutchison, J.S. & Zeleznik, A.J. (1984). The rhesus monkey corpus luteum is dependent on pituitary gonadotrophin secretion throughout the luteal phase of the menstrual cycle. *Endocrinology* **115:** 1780–6.

Jarvis, J.U.M. (1981). Eusociality in a mammal: Co-operative breeding in naked mole-rat colonies. *Science, N.Y.* **212:** 571–3.

Jarvis, J.U.M. (1984). African mole-rats. In *The encyclopedia of mammals* **2:** 708–11. (Ed. Macdonald, D.). George Allen and Unwin, London.

Jarvis, J.U.M. (In press). Reproduction. In *The biology of the naked mole-rat.* (Eds Jarvis, J.U.M., Sherman, P.W. & Alexander, R.D.). Cambridge University Press, Cambridge.

Karsch, F.J., Bittman, E.L., Foster, D.L., Goodman, R.L., Legan, S.J. & Robinson, J.E. (1984). Neuroendocrine basis of seasonal reproduction. *Rec. Prog. Horm. Res.* **40:** 185–232.

Kayanja, F.I.B. & Jarvis, J.U.M. (1971). Histological observations on the ovary, oviduct and uterus of the naked mole-rat. *Z. Säugetierk.* **36:** 114–21.

Kesner, J.S., Kaufman, J-M., Wilson, R.C., Kuroda, G. & Knobil, E. (1986). The effect of morphine on the electro-physiological activity of the hypothalamic luteinizing hormone-releasing hormone pulse generator in the rhesus monkey. *Neuroendocrinology* **43:** 686–8.

Keverne, E.B. (1979). Sexual and aggressive behaviour in social groups of talapoin monkeys. *Ciba Found. Symp.* (N.S.) **62:** 271–97.

Keverne, E.B. (1983). Pheromonal influences on the endocrine regulation of reproduction. *Trends Neurosci.* **6:** 381–4.

Kleiman, D.G. (1977). Monogamy in mammals. *Q. Rev. Biol.* **52:** 39–69.

Konner, M. & Worthman, C. (1980). Nursing frequency, gonadal function, and birth spacing among !Kung hunter-gatherers. *Science, N.Y.* **207:** 788–91.

Lacey, E.A. & Sherman, P.W. (In press). Social organization of *Heterocephalus glaber* colonies: evidence for a division of labour. In *The biology of the naked mole-rat.* (Eds Jarvis, J.U.M., Sherman, P.W. & Alexander, R.D.). Cambridge University Press, Cambridge.

Lee, P.C. (1987). Nutrition, fertility and maternal investment in primates. *J. Zool., Lond.* **213:** 409–22.

Legan, S.J. & Karsch, F.J. (1980). Photoperiodic control of seasonal breeding in ewes: modulation of the negative feedback action of oestradiol. *Biol. Reprod.* **23:** 1061–8.

Leyendecker, G. & Wildt, L. (1983). Induction of ovulation with chronic intermittent (pulsatile) administration of Gn-RH in women with hypothalamic amenorrhea. *J. Reprod. Fert.* **68:** 397–409.

Liddell, C.E., Abbott, D.H., Faulkes, C.G., George, L.M. & Jarvis, J.U.M. (In press). Hormones and behaviour in captive colonies of naked mole-rats (*Heterocephalus glaber*): preliminary results. In *The biology of the naked mole-rat.* (Eds Jarvis, J.U.M., Sherman, P.W. & Alexander, R.A.). Cambridge University Press, Cambridge.

Lincoln, G.A. & Short, R.V. (1980). Seasonal breeding: nature's contraceptive. *Rec. Prog. Horm. Res.* **36:** 1–52.

Lloyd, J.A. & Christian, J.J. (1969). Reproductive activity of individual females in three experimental freely growing populations of house mice (*Mus musculus*). *J. Mammal.* **50:** 49–59.

McNeilly, A.S. (1979). Effects of lactation on fertility. *Br. med. Bull.* **35:** 151–4.

McNeilly, A.S. (1984). Prolactin and ovarian function. In *Neuroendocrine perspectives* **3:** 279–316. (Eds Miller, E.E. & Macleod, R.M.). Elsevier Scientific Publishing, Amsterdam.

McNeilly, A.S. (In press a). Lactation and the return of ovulation. In *Natural human fertility: Social and biological mechanisms.* The Eugenics Society, London.

McNeilly, A.S. (In press b). Suckling and the control of gonadotrophin secretion. In *The physiology of reproduction.* (Eds Knobil, E. & Neill, J.A.). Raven Press, New York.

McNeilly, A.S., Glasier, A. & Howie, P.W. (1985). Endocrine control of lactational infertility. In *Maternal nutrition and lactational infertility*: 1–16. (Ed. Dobbing, J.). Vevey/Raven Press, New York.

Malcolm, J.R. & Marten, K. (1982). Natural selection and the communal rearing of pups in African wild dogs (*Lycaon pictus*). *Behav. Ecol. Sociobiol.* **10:** 1–13.

Martensz, N.D., Vellucci, S.V., Herbert, J. & Keverne, E.B. (1986). β-endorphin in the cerebrospinal fluid of male talapoin monkeys related to dominance status in social groups. *Neuroscience* **18:** 651–8.

Massey, A. & Vandenbergh, J.G. (1980). Puberty delay by a urinary cue from female house mice in feral populations. *Science, N.Y.* **209:** 821–2.

Mattioli, M., Conte, F., Graleati, G. & Seren, E. (1986). Effect of naloxone on plasma concentrations of prolactin and LH in lactating sows. *J. Reprod. Fert.* **76:** 167–73.

Mech, D.L. (1970). *The wolf: The ecology and behaviour of an endangered species.* The Natural History Press, New York.

Meyer, S.L. & Goodman, R.L. (1986). Separate neural systems mediate the steroid-dependent and steroid-independent suppression of tonic luteinizing hormone secretion in the anoestrous ewe. *Biol. Reprod.* **35:** 562–71.

Moehlman, P.D. (1983). Socioecology of silver-backed and golden jackals (*Canis mesomelas* and *Canis aureus*). *Spec. Publs Am. Soc. Mammal.* No. 7: 423–53.

Nicholls, T.J., Goldsmith, A.R. & Dawson, A. (1984). Photorefractoriness in European starlings: associated hypothalamic changes and the involvement of thyroid hormones and prolactin. *J. exp. Zool.* **232:** 567–72.

Quigley, M.E., Sheehan, K.L., Casper, R.F. & Yen, S.S.C. (1980). Evidence for an increased dopaminergic and opioid activity in patients with hypothalamic hypogonadotrophic amenorrhoea. *J. clin. Endocrin. Metab.* **50:** 949–54.

Quigley, M.E. & Yen, S.S.C. (1980). The role of endogenous opiates on LH secretion during the menstrual cycle. *J. clin. Endocrin. Metab.* **51:** 179–81.

Roberts, A.C., Martensz, N.D., Hastings, M.H. & Herbert, J. (1985). Changes in the photoperiod alter the daily rhythm of pineal melatonin content, hypothalamic β-endorphin content and the LH response to naloxone in the male Syrian hamster. *Endocrinology* **117:** 141–8.

Robinson, J.E., Radford, M.H. & Karsch, F.J. (1985). Seasonal changes in pulsatile luteinizing hormone (LH) secretion in the ewe: relationship of frequency of LH pulses to day length and response to estradiol negative feedback. *Biol. Reprod.* **33:** 324–34.

Rood, J.P. (1980). Mating relations and breeding suppression in the dwarf mongoose. *Anim. Behav.* **28:** 143–50.

Ropert, J.F., Quigley, M.E. & Yen, S.S.C. (1981). Endogenous opiates modulate pulsatile luteinizing hormone release in humans. *J. clin. Endocrin. Metab.* **52:** 583–5.

Ruiz de Elvira, M-C. & Abbott, D.H. (1986). A backpack system for long-term osmotic minipump infusions into unrestrained marmoset monkeys. *Lab. Anim.* **20:** 329–34.

Schallenberger, E., Richardson, D.W. & Knobil, E. (1981). Role of prolactin in the lactational amenorrhoea of the rhesus monkey (*Macaca mulatta*). *Biol. Reprod.* **25:** 370–4.

Schilling, A., Perret, M. & Predine, J. (1984). Sexual inhibition in a prosimian primate: a pheromone-like effect. *J. Endocr.* **102:** 143–51.

Seal, U.S., Plotka, E.D., Packard, J.M. & Mech, L.D. (1979). Endocrine correlates of reproduction in the wolf. I. Serum progesterone, estradiol and LH during the estrous cycle. *Biol. Reprod.* **21:** 1057–66.

Short, R.V. (1976). The evolution of human reproduction. *Proc. R. Soc.* (B) **195:** 3–24.

Sirinathsinghji, D.J.S. & Martini, L. (1984). Effects of bromocriptine and naloxone on plasma levels of prolactin, LH and FSH during suckling in the female rat: responses to gonadotrophin releasing hormone. *J. Endocr.* **100:** 175–82.

Stevenson, M.F. & Rylands, A.B. (In press). The marmoset monkeys, genus *Callithrix*. In *Ecology and behaviour of neotropical primates* 2. (Eds Mittermeier, R. & Coimbra-Filho, A.F.). Academia Brasileira de Ciencias, Rio de Janeiro.

Summers, P.M., Wennink, C.J. & Hodges, J.K. (1985). Cloprostenol-induced luteolysis in the marmoset monkey (*Callithrix jacchus*). *J. Reprod. Fert.* **73:** 133–8.

Terborgh, J. & Goldizen, A.W. (1985). On the mating system of the cooperatively breeding saddle-backed tamarin (*Saguinus fuscicollis*). *Behav. Ecol. Sociobiol.* **16:** 293–9.

Van Vugt, D.A., Bakst, G., Dyrenfurth, I. & Ferin, M. (1983). Naloxone stimulation of luteinizing hormone secretion in the female monkey: influence of endocrine and experimental conditions. *Endocrinology* **113:** 1858–64.

Van Vugt, D.A., Lam, N.Y. & Ferin, M. (1984). Reduced frequency of pulsatile luteinizing hormone secretion in the luteal phase of the rhesus monkey: involvement of endogenous opiates. *Endocrinology* **115:** 1095–101.

Wasser, S.K. & Barash, D.P. (1983). Reproductive suppression among female mammals: implications for biomedicine and sexual selection theory. *Q. Rev. Biol.* **58:** 513–38.

Whitten, W.K. (1959). Occurrence of anoestrus in mice caged in groups. *J. Endocr.* **18:** 102–7.

Yen, S.S.C. (1978). The human menstrual cycle (integrative function of the hypothalamic-pituitary-ovarian-endometrial axis). In *Reproductive endocrinology*: 152–70. (Eds Yen, S.S.C. & Jaffe, R.B.). W. B. Saunders, Philadelphia.

Yen, S.S.C. (1986). Chronic anovulation due to CNS-hypothalamic-pituitary dysfunction. In *Reproductive endocrinology. Physiology, pathophysiology and clinical management* (2nd edn): 500–45. (Eds Yen, S.S.C. & Jaffe, R.B.). W. B. Saunders, Philadelphia.

Symp. zool. Soc. Lond. (1988) No. 60 29–37

Assessment of the effects of chemicals on gamete production in captive and wild animals

A. J. HOLLOWAY
and H. D. M. MOORE

MRC/AFRC Comparative Physiology Group
Institute of Zoology
The Zoological Society of London
Regent's Park
London NW1 4RY, UK

Synopsis

Industrial and environmental chemicals may disrupt reproduction in animals by inhibiting gamete production and function. In this short review several examples are given, including the effects of dibromochloropropane on human testis function, the action of polychlorinated biphenols on Baltic ring seal reproduction, and the effects of DDE pesticide on avian eggs.

In order to evaluate accurately the effects of toxicants on gamete production in animals, sensitive assay systems are required. These must be both physiologically and ecologically relevant and economical. Tests range from the manipulation and fertilization *in vitro* of gametes recovered from laboratory animals exposed to toxicants, to the limited sampling of wild populations at risk. The usefulness of such tests to assess reproductive toxicity of chemicals is discussed.

Introduction

Environmental pollution with manufactured chemical compounds is continuously increasing. Some agents such as herbicides and pesticides are released deliberately, but others such as sulphur dioxide and methyl mercury are the waste products of industrial processes. Inevitably, in high doses some of these chemicals will be toxic. If reproductive processes were to be disrupted, whole populations would be endangered. Because man, domesticated animals and wild animals are all at risk there are strong economic and ethical arguments in favour of identifying dangerous industrial and other environmental chemicals and taking the necessary precautions to minimize their effects.

To establish whether those engaged in the manufacture or use of chemicals and drugs are at risk, toxicity is usually evaluated in laboratory mammals. However, assessment of the risk to man by extrapolation has

ZOOLOGICAL SYMPOSIUM No. 60
ISBN 0-19-854007-8

sometimes been shown to be unreliable. For example, dibromochloropropane (DBCP) which was widely used as a nematocide from the mid 1950s to the mid 1970s, caused infertility in male employees engaged in its manufacture. Toxicity testing in rabbits, rats and guinea pigs had shown that DBCP caused a reduction in testicular weight and, in the rat, a reduction in sperm numbers coupled with an increase in the proportion that were morphologically abnormal (Torkelson *et al.* 1961). These effects were considered to be minimal and, beyond the recommendation that the concentration of DBCP in air should be kept below 1 ppm, no further precautions were deemed necessary. In retrospect, there were clear indications that DBCP was a testicular toxicant but the estimation of the risk to man proved to be inadequate.

The persistence of toxicants in the environment and their eventual distribution throughout food chains means that animals may be affected indirectly. Polychlorinated biphenols (PCBs), used in capacitators and transformers and also as hydraulic and heat exchange fluids, are widespread water contaminants which persist in food chains. In the Baltic sea they are found in significant concentrations in herrings and hence in ring seals. Only 27% of adult female Baltic ring seals are pregnant at any one time compared with 80–90% in uncontaminated waters. Experimental evidence suggests that PCBs act by causing fetal death. Until recently, another persistent toxicant was 1,1-dichloro-2-bis-parachlorophenyl-ethylene (DDE), which was used as a pesticide in developed agricultural systems. DDE causes birds of some species to produce eggs with thin shells. This has led to a major decline in the numbers of several raptors, of which the most notable is the peregrine falcon in North America and Europe (Hickey 1969; Newton 1979).

Perhaps of most concern is the general release of toxicants from numerous industrial processes. Acid precipitation brought about by the generation of sulphur dioxide in industrial societies has gained widespread publicity. The lowered pH of freshwater systems has severely affected the reproductive capacity of many species of fish (Beamish 1976). Damage to the integument and branchial epithelium of the young, both before and after hatching, impairs respiration and ion homeostasis, with catastrophic results.

It is important to assess, by population studies, the gross effects that toxicants have already had on ecosystems, but the extensive long-term field work needed to determine what changes have occurred in reproductive capacity may not always be feasible. Indirect methods of assessing reduced fertility will be required when direct observation is impossible. This laboratory has concentrated on assessing gamete production and quality, because they are central to the reproductive process (Fig. 1). Once the types of lesion expected to arise after exposure to certain chemicals and families of chemicals have been established in laboratory animals, wild animals at risk can be

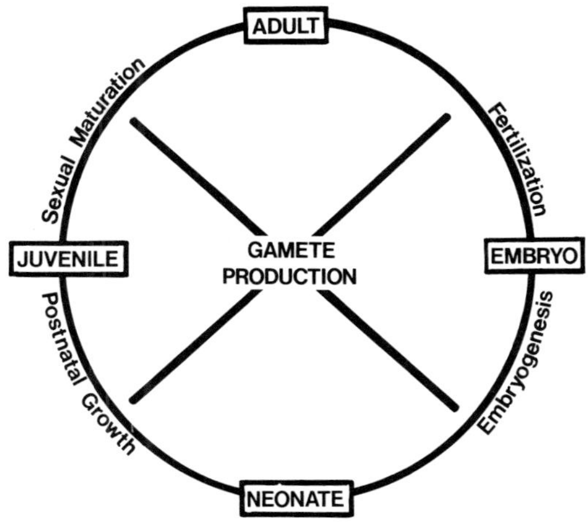

Fig. 1. A diagram illustrating the stages in the life cycle that play an integral role in the production of gametes.

screened for such lesions. Nevertheless, more effective screening of chemicals before their release into the environment is necessary. For this, the range of tests currently used in reproductive toxicity testing must be broadened to provide more information about the nature of the toxicity and to include more phyla. While not attempting to describe a comprehensive battery of tests for all phyla, we have selected a few examples that illustrate ways in which these issues may be tackled.

Assessment in mammals

Reproductive toxicity is assessed by the administration of a potential toxicant in relatively high doses and the use of histological examination and of multi-generation and serial breeding trials to determine its effect. The cell types affected are demonstrated histologically, but the high dose of the toxicant may make identification of the initial lesion impossible. Reduction in fertility caused by focal testicular damage is not always detectable by the currently used breeding trials. Serial breeding will provide limited information about the effects of the toxicant on libido, fertilization, dominant lethal mutations, and post-implantation embryonic development. This information, although essential, is not always sufficient for an adequate evaluation of potential hazards. For instance, it does not assist in distinguishing between fertilization failure and early embryonic loss.

Serial breeding trials with the usual laboratory animals (rat, mouse, rabbit, hamster) are relatively insensitive. The evolution of eutherian mammals has ensured considerable overproduction of spermatozoa. Under normal circumstances, each ejaculate contains a huge excess of sperm and the epididymal reserve is such that brief interruptions of spermatogenesis would probably go unnoticed. Fecundity may appear normal when only a small proportion of sperm remain viable and capable of fertilizing. This applies particularly to laboratory animals, which may have been selected for their fertility. Most laboratory animals are polytocous and are kept in conditions that ensure breeding throughout the year. Animal populations at risk from toxicants are unlikely to enjoy the optimal conditions under which laboratory species are kept, and will be more vulnerable as a result. Moreover, a temporary reduction in fertility at a critical time might have serious consequences in seasonally breeding animals. It is therefore necessary to devise assays of improved sensitivity, capable of providing more detailed information about the consequences of toxic exposure, so that the concentrations at which environmental toxicants become harmful may be accurately determined.

In order to overcome the problem of epididymal reserves and redundancy of spermatozoa in laboratory animals, it is desirable to determine by titration and to use in tests the minimum number of sperm required for maximum fertility. In this way, any reduction in the functional capacity of the gametes will become apparent. The rabbit has been used to provide such an assay system. Ejaculates were obtained from bucks by means of an artificial vagina and teaser doe; after washing and centrifugation the spermatozoa were resuspended in an appropriate medium and the density was adjusted to 10–30 million/ml. Follicular maturation was induced in does by the administration of serum gonadotrophin and 65 h later the animals were anaesthetized. Both the uterine horns were then exposed by a midline abdominal incision and ligated. Spermatozoa in known numbers were injected directly into each horn. Ovulation was then induced by chorionic gonadotrophin. The oocytes and embryos were recovered 24 h later, examined for fertilization and cultured, so that early embryonic development could be observed. The minimum number of spermatozoa required for maximum fertilization in the rabbit was 1 million per uterine horn (Fig. 2). This technique has been used to demonstrate changes in the fertilizing capacity of spermatozoa after exposure to doses of a testicular toxicant well below those judged harmless by serial breeding trials. Bucks were given a single threshold dose, by mouth, of ethylene glycol monomethyl ether (EGME), a testicular toxicant that specifically affects pachytene spermatocytes and has a lesser effect on leptotene and zygotene spermatocytes and late-stage spermatids. Lowered fertility was seen 5 weeks and 10–12 weeks after dosing (Fig. 3). Spermatozoa ejaculated 5 weeks after exposure to the toxicant would have been pachytene spermatocytes at the time of treatment and those ejaculated after 10–12 weeks would have been produced in the second spermatogenic cycle after exposure. This technique is much more sensitive than conventional

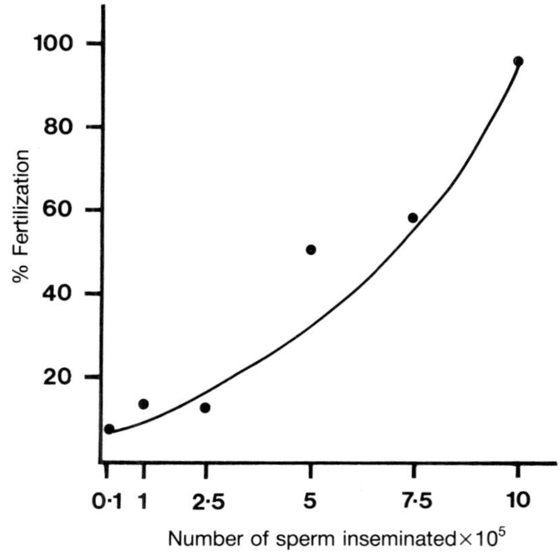

Fig. 2. A graph showing the relationship between the percentage of oocytes fertilized and the number of spermatozoa introduced into the uterus in the rabbit.

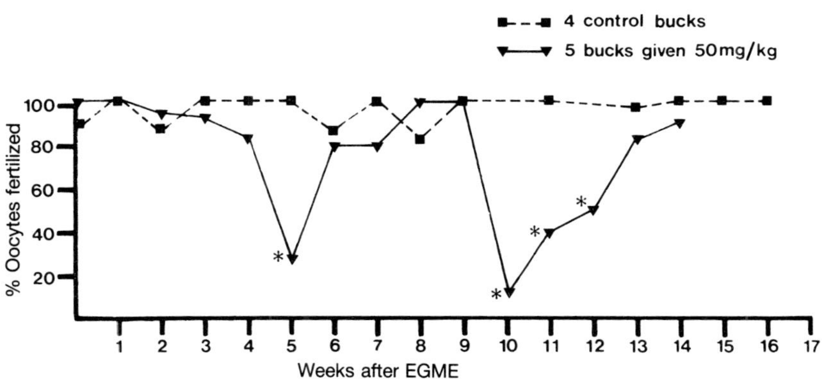

Fig. 3. A graph showing how the fertilizing ability of rabbit spermatozoa is affected when the males are treated with 50mg EGME/kg. *, Significantly different from control ($p < 0.05$).

methods used for detecting changes in the fertilizing capacity of spermatozoa.

In vitro fertilization (IVF) is the basis of another sensitive method of assessing the fertilizing capacity of spermatozoa. In rats such a method has proved more sensitive than the conventional breeding trial for detecting the effects of EGME (Fig. 4). As a result, safe levels of exposure can be determined with greater confidence than hitherto. Moreover, it is possible that,

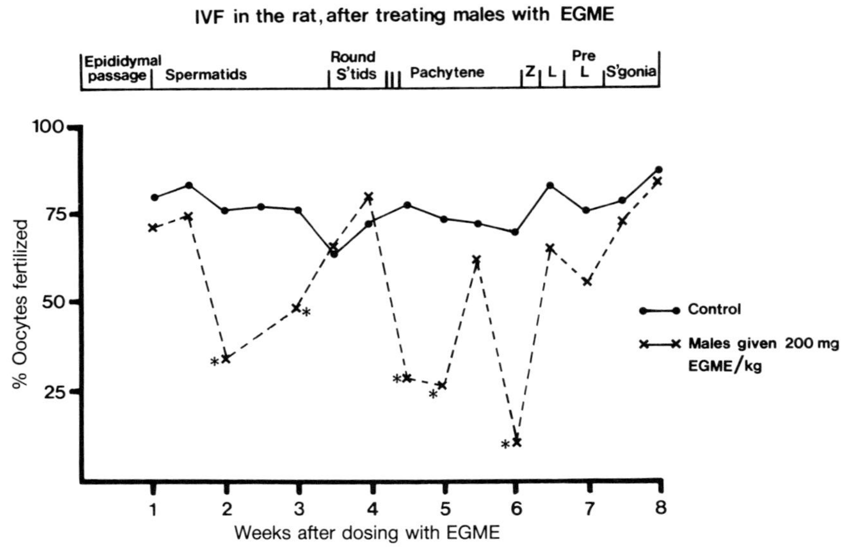

Fig. 4. A graph showing how the fertilizing ability of rat spermatozoa in an *in vitro* fertilization system is affected when the males are treated with 50mg EGME/kg. *, Significantly different from control ($p < 0.05$).

with increasing experience of such methods, reduction in *in vitro* fertilizing capacity will be found to be associated with morphological, physiological or biochemical abnormalities in spermatozoa. Information of this kind might be useful as the basis of indirect methods for screening wild populations.

Assessment in birds

The screening of birds for reproductive abnormalities caused by toxic chemicals presents a number of problems. It is necessary to use species that are readily available in large numbers and will breed in captivity. Birds that lay eggs repeatedly and have precocious young—for example domestic hens, ducks and quail—are particularly suitable for the rapid accumulation of data. The reproductive consequences of toxic exposure of the parents can be assessed by observing shell thickness, residues within eggs, hatching, growth and so on. Artificial incubation necessitates the use of a species whose young are precocious. Parental behaviour may be studied in the ring dove, which lays a single clutch of eggs and whose young are dependent for at least the first 14 days after hatching. It can be bred in small cages throughout the year and is a useful laboratory test species.

A simple and economical method of assessing the effects of toxicants is by their injection directly into the yolk sac of eggs during incubation (McLaughlin *et al.* 1963). This method was used to test the toxicity of a number of insecticides to chicken embryos (Dunachie & Fletcher 1969); the relative toxicities of the different compounds accorded with those found by other reproductive assays. The results of the yolk sac assay are, however, difficult to interpret and the potential of the method as a screening technique needs further critical evaluation.

Different avian species vary greatly in their susceptibility to toxic chemicals. Thus, whereas 20 ppm of Aroclor (a chlorinated hydrocarbon) causes almost complete reproductive failure in the chicken, 25 ppm has no effect on the mallard (Custer & Heinz 1980). It is therefore difficult to extrapolate from laboratory species to wild populations. If the specific effects of certain toxicants were known, for example shell thinning or metabolic or hormonal disturbance, screening of the wild populations at risk would be relatively simple and informative. Hormonal radio-immunoassays, which require only small volumes of blood, are a possible means of monitoring the effects of pollutants. Pollutant-induced changes in the concentrations of plasma sodium and circulating hormones have been demonstrated in a number of seabirds (Peakall, Hallett *et al.* 1980; Peakall, Tremblay *et al.* 1981). It is possible to measure average metabolic rate by the injection of doubly labelled water and to use the changes in the ratios of ^{3}H:H and ^{18}O:^{16}O in blood to calculate the rate (Lifson & McClintock 1966). In sampling it would be necessary to examine birds from unaffected as well as potentially affected populations, but if laboratory data had revealed a specific effect the number of birds sampled would not need to be large.

Assessment in fish

Toxicity testing is of particular importance in fish because they represent the upper trophic level of aquatic systems and so toxicants tend to accumulate in them. The selection of species suitable for use in tests must take account of the difficulties in obtaining and maintaining the fish, as well as their ecological significance. Not surprisingly, freshwater species are more often suitable than marine species. Various species of salmon and trout, and smaller fish such as the guppy and minnow (freshwater and marine) are often used.

It is important to be aware of (1) the stages in the life cycle that are likely to be exposed to a toxicant and (2) the relative sensitivities of the different stages. As with mammals and birds, toxicants may directly affect the gametes or the immature forms or they may act by reducing the reproductive capacity of the adult. Full life-cycle studies would be needed to assess effects in the adults, but some assays enable fertilization and early embryonic development to be investigated. In many fish the gametes are released directly into the aquatic environment to permit external fertilization and are thus directly exposed to water pollutants. The effect of such chemicals on fertilization can be readily assessed by exposing the gametes and subsequently counting the number of fertilized eggs. Toxic effects in immature forms can be studied in medaka eggs, which are transparent and so permit direct observation of embryonic development. They can be kept in small vials containing the test solutions (Stoss & Haines 1979). A method of determining mutagenic effects in male guppies has been described by Matthews, Favor & Crenshaw (1978); the results are based on abnormalities in embryonic development and behaviour.

Conclusions

Assay systems for fertility must be both economical and ecologically relevant. Since extreme caution must be exercised in the extrapolation of results from laboratory animals to wild populations, it is also necessary to employ indirect methods of assessing fertility and thus of monitoring the species at risk. Gamete production is central to the reproductive process and is susceptible to disruption throughout the life cycle. Examination of the gametes will therefore provide information concerning reproductive failure and also an indication of general toxicity.

References

Beamish, R.J. (1976). Acidification of lakes in Canada by acid precipitation and the resulting effects on fishes. *Water Air Soil Pollut.* **6:** 501–14.

Custer, T.W. & Heinz, G.H. (1980). Reproductive success and nest attentiveness of Mallard ducks fed Aroclor 1254. *Envir. Pollut. Ser. A Ecol. Biol.* **21:** 313–18.

Dunachie, J.F. & Fletcher, W.W. (1969). An investigation of the toxicity of insecticides to birds' eggs using the egg-injection technique. *Ann. appl. Biol.* **64:** 409–23.

Hickey, J.J. (Ed.) (1969). *Peregrine falcon populations—their biology and decline.* University of Wisconsin Press, Madison, Milwaukee & London.

Lifson, N. & McClintock, R. (1966). Theory and use of the turnover rates of body water for measuring energy and material balance. *J. theor. Biol.* **12:** 46–74.

Matthews, J.G., Favor, J.B. & Crenshaw, J.W. (1978). Dominant lethal effects of triethylenemelamine in the guppy *Poecilia reticulata. Mutat. Res.* **54:** 149–57.

McLaughlin, J., Jr., Marliac, J.-P., Vernett, M.J., Mutchler, M.K. & Fitzhugh, O.G. (1963). The injection of chemicals into the yolks of fertile eggs prior to incubation as a toxicity test. *Toxicol. appl. Pharmacol.* **5:** 760–71.

Newton, I. (1979). *Population ecology of raptors.* Buteo Books, Vermillion, S. Dakota.

Peakall, D.B., Hallett, D., Miller, D.S., Butler, R.G. & Kinter, W.B. (1980). Effects of ingested crude oil on black guillemots: a combined field and laboratory study. *Ambio* **9:** 28–30.

Peakall, D.B., Tremblay, J., Kinter, W.B., & Miller, D.S. (1981). Endocrine dysfunction in seabirds caused by ingested oil. *Envir. Res.* **24:** 6–14.

Stoss, F.W. & Haines, T.A. (1979). The effects of toluene on embryos and fry of the

Japanese medaka *Oryzias latipes* with a proposal for rapid determination of maximum acceptable toxicant concentration. *Envir. Pollut.* **20**: 39–148.

Torkelson, T.R., Sader, S.E., Rowe, V.K., Kodama, J.K., Anderson, H.H., Loquvan, G.S. & Hine, C.H. (1961). Toxicologic investigations of 1,2-Dibromo-3-chloropropane. *Toxicol. appl. Pharmacol.* **3**: 545–59.

Symp. zool. Soc. Lond. (1988) No. 60 39–80

Comparative quantitative studies of growth and reproduction

R. D. MARTIN

Anthropologisches Institut
Universität Zürich-Irchel
Winterthurerstrasse 190
8057 Zürich, Switzerland

and A.M. MacLARNON

Department of Anthropology
University College London
Gower Street
London WC1E 6BT, UK

Synopsis

Comparisons of reproductive parameters between mammals of different body weights are hindered by the confounding effects of scaling to body size. Because numerous parameters scale in a regular but non-linear fashion with body size, it is necessary to use appropriate techniques to eliminate scaling effects in order to conduct meaningful comparisons between species. Recent advances in the application of bivariate allometric analyses to interspecific comparisons have permitted new insights into numerous biological parameters, ranging from physiological measures to ecological characteristics. This contribution presents the results of allometric analyses conducted on gestation period and neonatal body weight in a large sample of placental mammals, with special emphasis on primates.

Comparative analyses of reproductive parameters are further complicated by the fact that there are differences between groups of mammals in the state of development of the neonate, as reflected in the classical distinction between 'altricial' and 'precocial' mammals. In the course of establishing a standard scaling relationship between gestation period and maternal body weight for placental mammals, it is demonstrated that a clear distinction can be drawn between two groups of mammals. One group (broadly equivalent to altricial mammals) is characterized by large litters of poorly developed offspring, born after a relatively short gestation period (e.g. insectivores, carnivores, myomorph rodents, sciuromorph rodents). The other group (broadly equivalent to precocial mammals) is characterized by small litters of well developed offspring, born after a relatively long gestation period (e.g. artiodactyls, perissodactyls, pinnipeds, primates and cetaceans). As a direct consequence of differences in gestation period, a similar distinction applies in the scaling of neonatal body weight to maternal body size, though the separation is not as clear as with gestation period because the proportional allocation of maternal resources to fetal growth is greater in altricial mammals than in precocial mammals. It is shown that the overall scaling of neonatal weight to maternal weight in placental mammals can be expressed in a combined formula that takes account of gestation period and the

ZOOLOGICAL SYMPOSIUM No. 60
ISBN 0-19-854007-8

rate of allocation of maternal resources (as indicated by the ratio between neonatal body weight and gestation period).

Through examination of residual values for neonatal body weight in primates, it is shown that the relatively smaller neonates of strepsirhine primates (lemurs and lorises) cannot be explained satisfactorily as a direct consequence of their non-invasive, epitheliochorial placentation. Haplorhine primates (tarsiers, monkeys, apes and man), which possess invasive, haemochorial placentation, do indeed typically produce larger neonates than strepsirhine primates (lemurs and lorises). However, other placental mammals with non-invasive placentation (hoofed mammals; cetaceans) are found to produce the largest neonates, relative to maternal size, among placental mammals generally. Further, when gestation period is taken into account, it is found that the dwarf lemur (*Cheirogaleus medius*) in fact exhibits the highest relative rate of commitment of maternal resources to fetal development among the primates, despite its non-invasive placentation. Residual values for rate of maternal investment in fetal development and for basal metabolic rate in primates are closely correlated. Hence, the difference between strepsirhine primates and haplorhine primates in neonatal body weights, relative to maternal weight, can be ascribed at least in part to differences in metabolic rate relative to body weight. For mammals generally, it would seem that the metabolic turnover of the mother provides a major constraint on fetal development, both directly and indirectly.

Introduction

In recent years, scaling of biological parameters to body size has received greatly increased attention. Several major books dealing specifically with this topic have appeared within a short space of time (Calder 1984; Jungers 1985; MacMahon & Bonner 1983; Peters 1983; Schmidt-Nielsen 1984). These books have considerably widened the scope of scaling studies and they reflect an increasing concern with the ubiquitous problems of scaling in biology.

There is an obvious need to develop reliable methods for comparison of developmental stages between mammal species, such that some kind of biological equivalence is ensured. It is, for example, misleading to compare neonates among different mammalian species if the time of development prior to birth (viz. the gestation period) is not comparable. This applies, for example, to the interpretation of data from laboratory mammals used as models for human developmental biology (Rudder 1978). A thorny problem arises in comparisons between mammalian species of different body sizes in that development times (such as gestation period) will themselves be related both to specific developmental processes and to the body size of the mother. Numerous studies have shown that many aspects of biological timing scale in a fairly predictable fashion with body weight (e.g. see Calder 1984), so—other things being equal—it is only to be expected that reproductive parameters such as gestation period will change in a fairly regular fashion with maternal body weight. Comparative studies of reproductive parameters between species therefore require a suitable technique for 'removing' the effects of body size.

Various studies made by means of allometric analysis have shown that it is possible to take account of body size effects, to a large degree at least, and to achieve more meaningful comparisons between species. This has been achieved in a variety of fields including cranial morphology, brain size, gastrointestinal dimensions, fetal development and adaptations for lactation (e.g. see Martin 1980, 1981, 1982, 1983, 1984a,b; Martin & Harvey 1985; Martin & MacLarnon 1985; Martin, Chivers, MacLarnon & Hladik 1985; MacLarnon, Chivers & Martin 1986; MacLarnon, Martin, Chivers & Hladik 1986). Following Huxley (1932), the approach is founded upon the general allometric formula $Y = k.\ X^{\alpha}$ relating a given biological parameter (Y) to some measure of overall body size (X). The two constants, the allometric exponent (α) and the allometric coefficient (k), determine the specific nature of the scaling relationship in any given case. In practice, it is customary to convert the allometric formula into logarithmic form, since the relationship then becomes linear: $\log Y = \alpha.\ \log X + \log k$. By determining an appropriate best-fit line for an empirical data set converted to logarithmic form, it is possible to establish the values of α and k for a given scaling relationship.

Huxley (1932) was largely concerned with intraspecific scaling relationships operating during development and this is an area in which work has naturally continued. However, the focus of many recent investigations, including the present study, is primarily upon interspecific comparisons and the procedure of allometric analysis will therefore be explained in this context. Here, each point in an analysis represents the average condition for a given biological parameter (Y) and for body weight (X) of an entire species. In a simple case, there will be a single, clearly identifiable trend in the bivariate logarithmic plot (e.g. as in the set of circles clustered around the lower best-fit line in Fig. 1A). The best-fit line will therefore provide a good indication of the general trend, which may be referred to as the scaling principle. The scaling principle is essentially reflected in the value of the allometric exponent. In such a simple case, there will be relatively little scatter of points around the best-fit line, but some information may be extracted from the pattern of deviation of points from the ideal relationship. Species that lie above the line have larger than expected values for the given biological parameter, in relation to their body sizes, whereas species that lie below the line have values smaller than expected. The distances of individual points from the line (viz. their residual values) indicate the extent to which the special adaptations of the species deviate from the common scaling trend. Individual species with extreme adaptations are recognizable as outliers, points noticeably distant from the best-fit line.

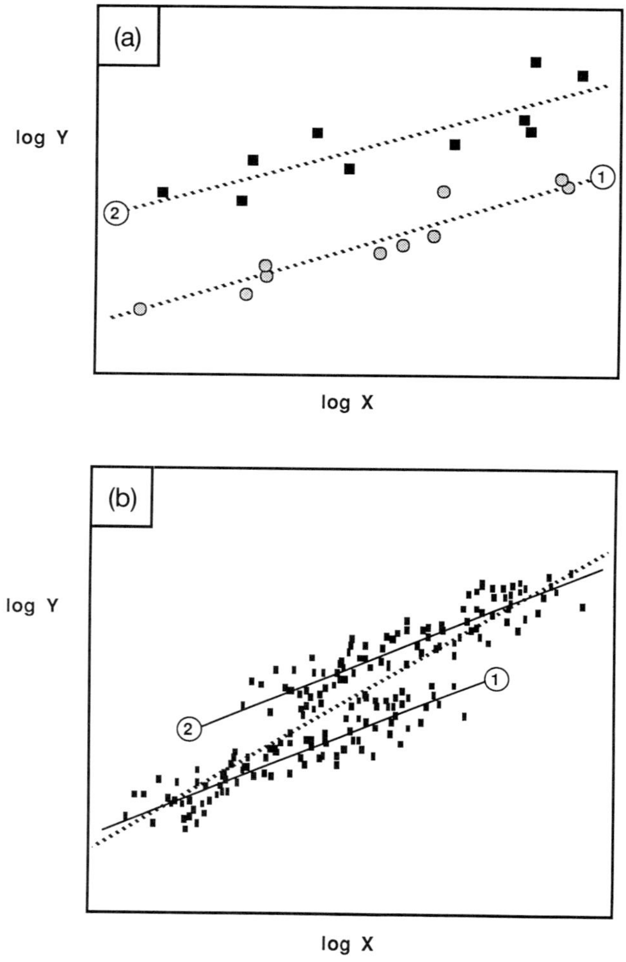

Fig. 1. Illustration of the basic principles of allometric analysis as applied in this paper. In a relatively simple situation (a), data for species from two different taxonomic groups (circles *versus* squares) have been plotted on logarithmic co-ordinates and a separate best-fit line has been fitted to each group (dotted lines 1 and 2). Both groups follow the same scaling principle, so the best-fit lines have closely similar slopes (= allometric exponent values). However, values for the allometric coefficient (k) are markedly different for the two groups, so the two lines are separated by a substantial vertical distance (= 'grade shift'). Within each group, there is some scatter about the best-fit line ($r = 0.96$ in both cases). Departures of individual species from the idealized scaling principle represented by a best-fit line are indicated by their vertical distances from the line (= residual values). In (b), a more complex situation is shown in which there is close approximation between two allometric grades represented by large samples of data. A best-fit line determined for the entire data set (dashed line) has a strikingly higher slope than individual best-fit lines (1 and 2) determined separately for the two sub-sets (grades). If the subdivision of the data set is not recognized, misleading residual values will be calculated relative to the overall best-fit line, resulting in 'grade confusion'.

Determination of the best-fit line for a given data set presents some statistical problems that have not yet been clearly resolved (see Harvey & Mace 1982 for a valuable discussion of several relevant issues). If the fit of the data to the best-fit line is very good, it matters little which particular line-fitting technique is used. However, if the data exhibit a moderate to large degree of scatter, reflected in relatively low values of the correlation coefficient (r), the question of choice of an appropriate best-fit line becomes critical. Most authors have simply used regressions for bivariate allometric analyses, but there are good reasons to question the use of such an asymmetrical line-fitting approach for the analysis of biological parameters. Use of a regression of Y on X implies that Y is definitely dependent upon X and that X is measured without error. In a case where two biological parameters are plotted against one another (e.g. gestation period against body weight), these conditions do not clearly apply. It is therefore preferable to use a symmetrical line-fitting technique that does not require an *a priori* distinction between dependent and independent variables. The major axis is a suitable, symmetrically determined best-fit line and has been used for all empirically determined scaling relationships reported below. Given the unresolved problems attached to the choice of an appropriate empirical best-fit line, it is advantageous, wherever possible, to fit a line of fixed slope to the data on the basis of prior theoretical considerations. Examples of this are provided.

Frequently, matters are even more complicated in that a given data set may include two or more fairly distinct subsets (Fig. 1b). Commonly it is found that separate lines of approximately the same slope may be fitted to data from different groups of animals. In other words, each group follows the standard scaling principle (reflected in closely similar values for the allometric exponent α), but there is a difference between groups in the allometric coefficient (k) and hence in the vertical alignment of the best-fit lines. Separate scaling relationships in different groups of mammals can be referred to as 'allometric grades' (Martin 1980; see Fig. 1). The possible existence of separate grades within a given data set further complicates the problems of line-fitting, since a single best-fit line applied to the entire data set will commonly have a slope that does not reflect the standard scaling principle operating within groups of functionally comparable animals. That is to say, there is a potential problem of 'grade confusion' in any allometric analysis (Fig. 1b). Accordingly, the analysis of any bivariate data set requires careful attention to the possible presence of distinct subsets as well as the application of appropriate line-fitting techniques. Objective recognition of the existence of 'grades' within a data set remains one of the major problems of allometric analysis and one possible solution, applicable in at least some cases, is discussed below.

Methods

The analyses reported below are based on a large data bank of mammalian body weights, brain weights and reproductive parameters assembled for the purposes of a comparative study of primate development. Most of the data were collated from previous studies, especially from those of Rudder (1979; pers. comm.) and Eisenberg (1981), with careful sifting and checking in each case; but new data (especially on neonate weights) were also derived from zoological collections. Results from a preliminary analysis of the scaling of gestation period in a sample of 394 placental mammals and of the scaling of neonatal body weight in a sample of 345 placental mammals have already been reported by Martin & MacLarnon (1985), and these results have been incorporated into the following text. For the additional analyses presented in this paper, the sample was reduced to 312 placental mammal species for which data were available on both gestation period and neonatal body weight, in order to ensure strict comparability. For gestation periods, the effects of delayed implantation were excluded wherever accurate information was available in the literature.

The following standard abbreviations are used throughout the text for reference to the main biological parameters associated with reproduction:

G = gestation period (days)
t = time (days)
W_m = maternal body weight (g)
W_n = individual neonatal body weight (g)
W_f = individual fetal body weight (g)
L = average litter-size.

Scaling of reproductive variables

Gestation period

The scaling of gestation period in placental mammals has been the subject of several broad comparative investigations (e.g. Blueweiss, Fox, Kudzma, Nakashima, Peters & Sams 1978; Kihlström 1972; Millar 1981; Robbins & Robbins 1979; Sacher & Staffeldt 1974; Western 1979; Western & Ssemakula 1982). In studies covering placental mammals generally, a scaling exponent value in the region of 0.25 has been reported (e.g. Blueweiss *et al.* 1978; Millar 1981; Sacher & Staffeldt 1974; see also summary tables in Calder 1984 and Peters 1983). Hence, it might seem that a general equation of the following form might be applicable to the scaling of gestation period:

$$G = k.W_m^{0.25} \tag{1}$$

In some cases, this particular value of the scaling exponent (viz. 0.25) has been cited as being significant within a framework of biological timing generally (e.g. Calder 1984). However, when individual orders of placental mammals are considered, a markedly lower exponent value is obtained for the scaling of gestation periods (Calder 1982; Kihlström 1972; Western 1979). The average exponent value for eight orders of placental mammals listed by Peters (1983) is 0.15, while Martin & MacLarnon (1985) found exponent values of 0.13, 0.13 and 0.11 for large samples of Primates, Artiodactyla and Carnivora, respectively. This in itself raises the possibility that some kind of grade confusion may be involved in allometric analyses of mammalian gestation periods. It is therefore essential to examine the scaling of gestation period among placental mammals more closely.

An obvious factor that must be considered with respect to the scaling of gestation periods is the state of development of the offspring at birth. Numerous studies, most notably those of Portmann (1938, 1939, 1941, 1962, 1965), have shown that a fundamental distinction should be made between altricial and precocial mammals (see also Martin 1975, 1983, 1984a). Altricial offspring are relatively poorly developed at birth and are typically born in multiple litters, whereas single births are the rule for precocial offspring, which are conspicuously well developed at birth. Useful practical indicators of the degree of development of a mammal at birth are: time of eye-opening and ear-opening (themselves indirect indicators of the degree of overall maturation of the nervous system); stage of development of the fur; litter size. Obviously, other things being equal, poorly developed offspring require a shorter gestation than well developed offspring. Equally obviously, there is likely to be a general inverse relationship between litter-size and gestation period across species, since the limited capacity of the uterus of each species of a given body size will require that the maximum neonatal size of offspring in multiple litters be smaller than the maximum neonatal size of a single offspring.

As has been shown by Portmann and others (e.g. see Portmann 1938, 1939, 1965), many large taxonomic groups of mammals are characterized by a single type of neonate. For instance, carnivores and insectivores (including tree-shrews, but excluding elephant-shrews) typically have altricial offspring, whereas hoofed mammals (artiodactyls and perissodactyls) and primates typically have precocial offspring. However, exceptions and intermediate cases occur which tend to obscure the distinction between altricial and precocial offspring. For instance, whereas most rodents (myomorphs and sciuromorphs) typically have altricial offspring, the hystricomorph rodents have offspring that tend to be precocial, though some species have been said not to fit either category clearly. Similarly, among the lagomorphs there are some species that have typical altricial offspring (e.g. rabbits and pikas), whereas others have offspring that conform more to the precocial pattern, at least in terms of state of development at birth (hares). Because of these exceptions and intermediate cases, it has often been implied that there is a smooth continuum between altricial and precocial offspring among mammals (e.g. Zeveloff & Boyce 1986) and even that the distinction between the two types of offspring is too blurred to be of practical value. However, this view is to a large extent attributable to confusion between the scaling of reproductive parameters (e.g. gestation period and neonatal body weight) to maternal body size and special adaptations of individual species that are only identifiable when scaling effects have been taken into account.

On the basis of Portmann's previously established criteria, Martin & MacLarnon (1985) divided a sample of 394 placental mammals into three groups (altricial, intermediate, precocial) by means of a dual criterion based on litter-size and time of eye opening. Definitely altricial mammals were those with an average litter-size of 3 or more and a minimum period of 5 days between birth and eye-opening. Definitely precocial mammals were those with an average litter-size of less than 1.5 and with eyes open by the time of birth. Species which fell between these two categories (i.e. those with an average litter-size between 1.5 and 3 and/or with eyes opening between birth and 5 days of age) were defined as 'intermediate'. Having made this tripartite distinction, empirical scaling relationships between gestation period and maternal body size were established for altricial and precocial mammals separately (Fig. 2). As is apparent from Fig. 2, altricial and precocial mammals constitute two distinct grades with relatively low scaling exponent values. For precocial mammals (line P), the exponent value is 0.11, while for altricial mammals (line A) the exponent value is 0.17 (Martin & MacLarnon 1985). When a single best-fit line is determined for the complete sample, including altricial, intermediate and precocial forms, an exponent value of 0.22 is obtained, close to the overall exponent value reported by other authors. Figure 2 shows quite clearly that this higher value is a product of grade confusion (see also Fig. 1b). The overall line (C) does indicate a general trend, in that there is a tendency for gestation periods to increase more rapidly with body size in placental mammals generally than is usually the case within individual orders of mammals. However, this overall trend results from the fact that the average body weight of precocial mammals is significantly greater than the average body weight of altricial mammals.

The empirical exponent values for precocial and altricial mammals determined from the two best-fit lines shown in Fig. 2 must also be examined more closely. The exponent value for altricial mammals (0.17) is greater than that for precocial mammals (0.11) and there is a possibility that further grade distinctions may exist, at least within the altricial category. The average exponent value determined by Martin & MacLarnon (1985) for five major taxonomic groups of placental mammals characterized by a specific

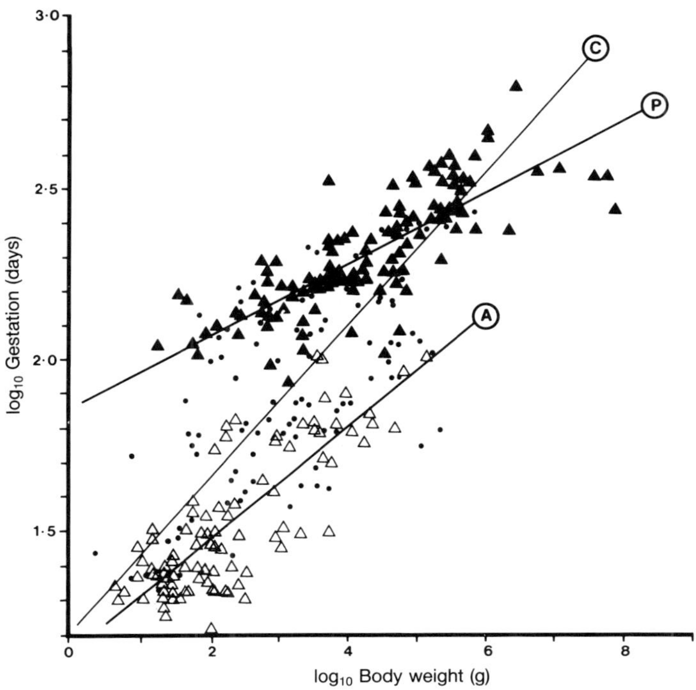

Fig. 2. Logarithmic plot of gestation period (days) against maternal body weight (g) for 394 placental mammal species. Best-fit lines (major axes) were determined separately for altricial mammals (open triangles; line A) and for precocial mammals (solid triangles; line P). An overall best-fit line (C) was also determined for the complete sample of altricial, precocial and intermediate species (the latter being indicated by small black dots). (Reprinted by permission from *Nature, Lond.* **313**: 220–223. Copyright © Macmillan Magazines Ltd.)

type of neonate (altricial: carnivores, myomorph rodents; precocial: artiodactyls, hystricomorph rodents, primates) was 0.12 (range: 0.09–0.13). This indicates that gestation period scales to body weight with a relatively low exponent value within individual mammalian groups characterized by a particular type of neonate. However, it is difficult to set objective limits to the *ad hoc* subdivision of a given sample into apparent 'grades' by the investigator and the danger exists that the process will be continued until—and only until—exponent values are obtained that fit the investigator's preconceptions. As was noted in the introduction, definition of a reliable criterion for the objective recognition of grade distinctions within a sample remains one of the major problems of allometric analysis. For this reason, a radically different approach was applied to the data set in Fig. 2.

Any bivariate data set that approximates to the bivariate normal distribution required by standard parametric line-fitting techniques (e.g. regression, major axis) is, by definition, a data set free of internal grade distinctions. If a line of *any* slope is fitted to the data set, passing through the standard reference point defined by the overall mean values for the two variables, the distribution of residual values should be consistently unimodal and approximately normal. By contrast, if a data set contains two well-defined sub-sets with parallel slopes (each sub-set approximating to the ideal requirement of a bivariate normal distribution), this will not apply (Fig. 1B). Indeed, if the two grades are fairly clearly separated, a line of the appropriate slope passed through the overall mean values for both variables should produce a distinct bimodal distribution of residual values. Therefore, successive fitting of lines of different slope to a given data set should provide an objective test for the existence of grades. Generation of a bimodal pattern within a certain range of slope values would indicate both the existence of two separate sub-sets of data and the approximate common slope of the best-fit lines for the sub-sets. The process remains a trial-and-error one, however, since it is difficult to establish a clear criterion for recognizing an optimal bimodal distribution of residual values derived from lines of closely similar slope.

By means of an iterative computer programme, lines of different slope were fitted to the data shown in Fig. 2 and a recognizable bimodal pattern of residuals was obtained for slope values between 0.17 and 0.07. Within this range, a slope value of 0.10 was taken as optimal, since it produced the clearest bimodal distribution with minimal misclassification of altricial and precocial mammals (Martin & MacLarnon 1985). This indicates that the appropriate allometric formula for the scaling of gestation periods within mammalian groups with comparable degrees of neonatal development is:

$$G = k.W_m{}^{0.10} \tag{2}$$

For the total sample of 312 placental mammal species analysed here, the formula for the overall line of fixed slope=0.10 is:

$$G = 38.8.W_m{}^{0.10} \tag{3}$$

The distribution of the residual values for altricial, intermediate and precocial mammals determined according to this reference formula is shown in Fig. 3. It can be seen that definitely altricial and definitely precocial mammals are almost completely separated and that the trough between the two peaks is located close to the logarithmic zero value. This indicates that relatively few species lie close to the overall line of slope 0.10; most lie substantially above or below this line (see also Fig. 2). Further, the residual values for species defined as intermediate do not obliterate the bimodal pattern shown by the species defined as altricial and precocial. Instead, they conform quite well to the existing bimodal pattern. Figure 3 therefore shows quite convincingly that all placental mammals in the sample can be sharply divided into two groups (which may be labelled 'altricial' and 'precocial' on

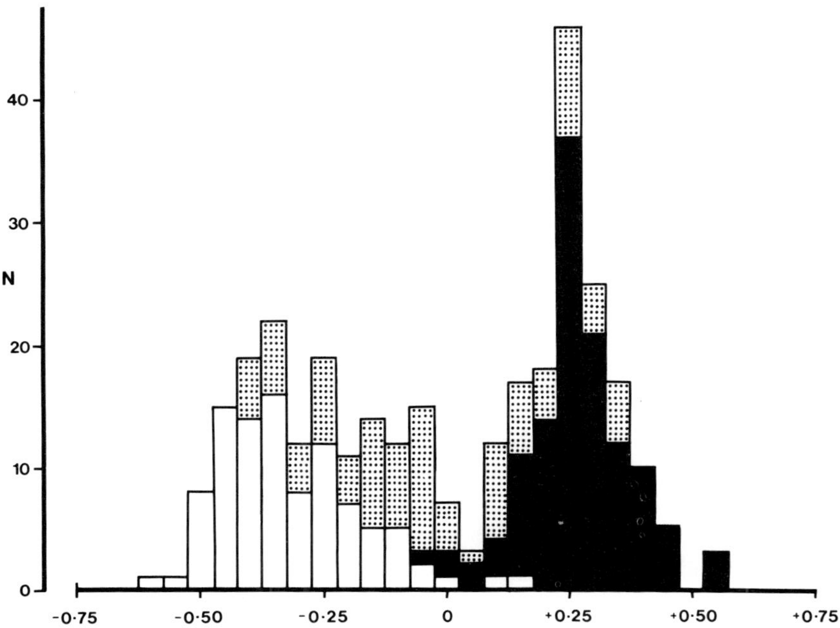

Fig. 3. Histograms of residual values for gestation periods, determined relative to a line of fixed slope 0.10, for the 394 placental mammal species indicated in Fig. 2. There is an almost complete separation between altricial mammals (white boxes) and precocial mammals (solid boxes). The inclusion of intermediate species (stippled boxes) does not obscure the bimodal pattern. (Reprinted by permission from *Nature, Lond.* **313**: 220–223. Copyright © Macmillan Magazines Ltd.)

the basis of an objectively recognizable distinction in gestation period relative to body size.

Figure 3 also shows that a substantial difference exists between altricial and precocial mammals in terms of gestation period relative to body size. The average values from the distributions of logarithmic residual values for altricial mammals (−0.32) and for precocial mammals (+0.27) are separated by a logarithmic interval of 0.59, indicating an approximate *fourfold* difference in gestation period between typical altricial and typical precocial mammals. In view of the essentially cubic relationship between fetal growth and time (see later), this difference is extremely large. Other things being equal, it would lead to typical precocial neonates being 64 times bigger than typical altricial neonates for any given maternal body weight!

Having established the validity of a fundamental distinction between 'altricial' and 'precocial' mammals, it is possible to proceed to a finer analysis of the scaling of gestation periods for individual orders of placental mammals. This can be achieved by examining distributions of residual values for these individual mammalian orders. The overall samples of placental mammals represented in Figs 2 and 3 included five orders for which sufficient numbers of species were documented to justify presentation in histogram form: Rodentia ($N = 110$), Insectivora, including tree-shrews ($N = 18$), Carnivora ($N = 46$), Primates ($N = 52$) and Artiodactyla ($N = 40$). Distributions of residual values within these five orders are shown in Fig. 4. It can be seen that the insectivores and carnivores, which are generally regarded as having altricial neonates, typically have negative residual values for their gestation periods. Incidentally, in view of continuing discussion about a possible phylogenetic relationship between tree-shrews and primates, it should be noted that the residual value for the common tree-shrew *Tupaia belangeri* (−0.16) lies in the middle of the range for insectivores (0.00 to −0.37) and is well below the lowest value recorded for any primate, namely for the dwarf lemur *Cheirogaleus medius* (−0.02). Relative to body size, *Cheirogaleus* has a gestation period 38% longer than that of *Tupaia*. Because of the approximately cubic relationship between fetal size and elapsed time (see later), *Cheirogaleus* is able to produce a neonate almost twice as large as that of *Tupaia*, despite a relatively lower level of maternal investment.

In contrast to insectivores and carnivores, the primates and artiodactyls, which characteristically have neonates of the precocial type, typically have positive residual values. The rodents, however, show an extended distribution with a just discernible bimodal pattern. Further analysis (Fig. 5) shows that members of the mouse suborder (Myomorpha) and of the squirrel suborder (Sciuromorpha), which are generally characterized by altricial offspring, typically have negative residual values. On the other hand, members of the porcupine suborder (Hystricomorpha), which characteristically have precocial offspring, typically have positive residual values. The pattern of residual values for rodents therefore confirms the special status of the hystricomorph rodents with respect to degree of neonatal development. In the case of the rodents, the distinction between altricial and precocial mammals is therefore clearly recognizable at the subordinal level, rather than at the ordinal level as with most other groups of placental mammals. It should be noted that the minimal residual value found among rodents, and among placental mammals generally, is that of −0.59 for the golden hamster (*Mesocricetus auratus*). The implication that this species therefore has the shortest gestation period, relative to body size, among placental mammals fits well with the fact that the golden hamster is widely used for studies of development of the central nervous system, since the neonatal brain exhibits the least advanced condition so far reported among placentals. In a similar vein, special mention should be made of the fact that the pig (*Sus scrofa*) is the only artiodactyl species to exhibit a negative residual value (−0.05) for its gestation period. Although piglets are born with their eyes open, unlike most other artiodactyls the wild pig constructs nests and has multiple litters

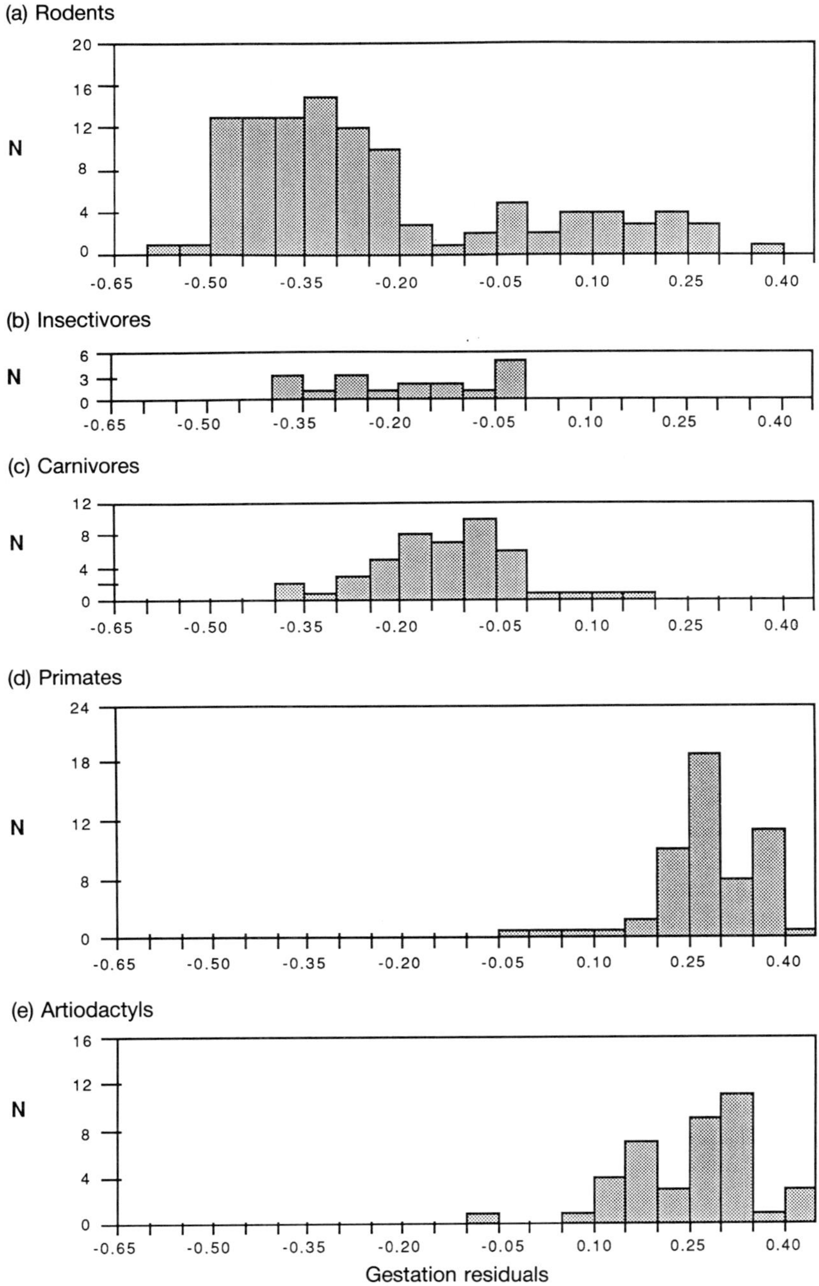

Fig. 4. Histograms of residual values for gestation periods, determined relative to a line of fixed slope 0.10, for five orders of placental mammals (Rodentia, Insectivora, Carnivora, Primates and Artiodactyla).

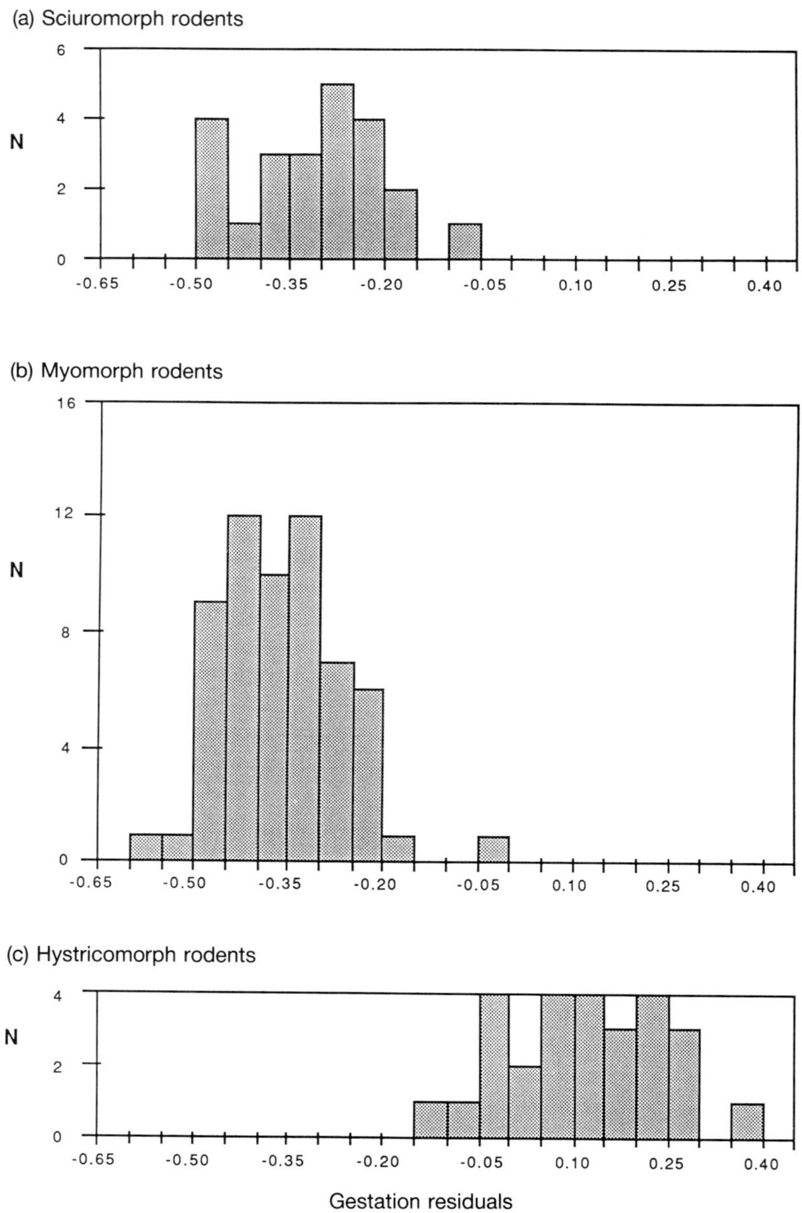

Fig. 5. Histograms of residual values for gestation periods, determined relative to a line of fixed slope 0.10, for the three suborders of Rodentia (Sciuromorpha, Myomorpha and Hystricomorpha). The altricial sciuromorphs and myomorphs show a pattern of distribution comparable to that of insectivores and carnivores, while the precocial hystricomorphs show a pattern similar to that of primates and artiodactyls.

(three to five neonates per litter according to Eisenberg 1981) and is therefore intermediate between typically altricial and typically precocial mammals. The closely related peccary (*Tayassu*) and warthog (*Phacochoerus*) also have relatively low residual values for their gestation periods (0.15 and 0.17, respectively), but they fall within the range of distribution for other artiodactyls.

For other groups of placental mammals, sample sizes are too small to warrant presentation of the residual values in histogram form in Fig. 4. However, some valuable additional points can be drawn from an assessment of the values for individual orders. The only order of mammals other than Insectivora and Carnivora to show consistently negative residual values is the Lagomorpha ($N = 7$; range of residual values: −0.23 to −0.42). This result is, in fact, somewhat unexpected. As mentioned above, although rabbits and pikas have typically altricial offspring, hares produce neonates that are quite well developed at birth, though multiple litters are the rule. (Hares would therefore be classified as 'intermediate' according to the criteria set out by Martin & MacLarnon (1985).) Nevertheless, hares closely resemble rabbits in having very short gestation periods in relation to body size. Therefore, the hares are unusual among all the placental mammals investigated in that a precocial level of neonatal development is associated with strikingly short gestation periods. Among the edentates (N=6), positive residual values are found as expected for most species (armadillos, 0.12 for *Dasypus* and 0.18 for *Tolypeutes;* sloths, 0.29 for *Bradypus* and 0.56 for *Choloepus*). However, in the two hairy armadillos negative residual values are found (*Euphractus* = −0.15; *Chaetophractus* = −0.13). Thus, once the scaling effects of body size have been taken into account, it emerges that *Dasypus* and *Tolypeutes* have gestation periods approximately twice as long as *Euphractus* and *Chaetophractus*. This represents an interesting anomaly that requires further investigation. The single representative of the order Tubulidentata, the aardvark *Orycteropus*, was originally found to have a barely positive residual value of 0.02 when the gestation period of 120 days cited by Eisenberg (1981) was taken. This would indicate that *Orycteropus* is intermediate between typically altricial and typically precocial mammals. However, *Orycteropus* typically has a single, well-developed neonate and clearly fits the precocial pattern, rather than the altricial pattern. In fact, checking of the original reference given by Eisenberg (viz. Melton 1976) revealed that the gestation period was cited as '7 months', presumably following Dorst & Dandelot (1970). When a gestation period of 210 days is taken instead, the residual value rises to 0.26, falling well within the range for typically precocial mammals. In this instance, the anomalous residual value originally obtained for *Orycteropus* led directly to recognition of a printing error.

All of the remaining mammalian orders investigated have positive residual values for their gestation periods, as would be expected from the fact that they generally have typically precocial offspring. The range of values for Chiroptera ($N = 6$) is 0.04 to 0.43, for Cetacea ($N = 4$) 0.06 to 0.45, for Perissodactyla ($N = 4$) 0.39 to 0.47, and for Pinnipedia ($N = 14$) 0.21 to 0.50. The two elephant species have the highest recorded mammalian values, 0.56 and 0.57, respectively, while the single representative of the Hyracoidea covered by the survey (*Procavia*) has a value of 0.38 and the single member of the Sirenia (*Trichechus*) has a value of 0.42. It should also be noted that among the cetaceans the only whale included in the sample (*Balaenoptera*) has a relatively low value of 0.06, whereas the three dolphins included have consistently high values of 0.43–0.45.

Neonatal body weight

A similar kind of analysis can be carried out for the scaling of neonatal body weight to maternal body weight. Martin & MacLarnon (1985) found that the average scaling exponent value for 345 placental mammal species was 0.93, very close to values reported by other authors for overall scaling of neonatal body weight in placental mammals (e.g. Leitch, Hytten & Billewicz 1959; Leutenegger 1976, 1977; Blueweiss *et al.* 1978; Millar 1981). As with gestation period, however, the determination of an appropriate scaling exponent is bedevilled by grade differences. In particular, it is only to be expected that precocial mammals, with their longer gestation periods relative to maternal body weight, should have larger neonates for a given maternal body weight than altricial mammals. The expected grade difference between precocial and altricial mammals with respect to neonatal weight (relative to maternal weight) is indeed found (see Fig. 4 in Martin 1984a). However, grade separation between altricial and precocial mammals is not so clear as with gestation periods, because, at any given body weight, mothers with altricial offspring typically devote a greater proportion of their resources per unit time to fetal growth than mothers with precocial offspring (see later). Hence, the iterative procedure used to determine the optimal exponent value for the scaling of gestation period does not yield clear results for the scaling of neonatal body weight. Using the definitions mentioned above, Martin & MacLarnon (1985) found that for precocial mammals ($N = 130$) the empirical exponent value was 0.91 and for altricial mammals ($N = 73$) 0.75. These two values yield an average value of 0.83, lower than the scaling exponent of 0.93 for all placental mammals taken together. The average scaling exponent value for five major groups of mammals each characterized by a single neonate type (primates, artiodactyls, carnivores, myomorph rodents, hystricomorph rodents) was found to be 0.81 (range: 0.64–0.89). There are also, however, a number of unusual species that are conspicuous outliers with respect to the general trend.

Among carnivores, for example, the bears and the giant panda produce unusually small neonates (see Martin 1984a). Since maternal body weight happens to be high in these species, the outlying points act to reduce the overall scaling exponent values both for carnivores specifically and for altricial mammals generally. It was therefore decided for the purposes of the present analysis to take a standard exponent value of 0.85 for the scaling of neonatal body weight in placental mammals and to calculate residual values relative to a line with a fixed slope of this value. The resulting equation has the following formula:

$$W_n = 0.16.W_m{}^{0.85} \quad (4)$$

At this point, it is possible to enrich the information obtained from interspecific allometric comparisons by inclusion of the results of intraspecific studies of the relationship between fetal body weight (W_f) and elapsed time (t). Empirical studies of fetal development as a function of time in individual mammal species (e.g. MacDowell & Allen 1927; MacDowell, Allen & MacDowell 1927; Huggett & Widdas 1951; Payne & Wheeler 1967a, 1968; Roberts & Perry 1974) have yielded a standard fetal growth formula relating fetal weight (W_f) to time elapsed (t), following an initial 'lag phase' (t_0):

$$W_f = a.(t - t_0)^3 \quad (5)$$

This formula, which can also be derived theoretically (Payne & Wheeler 1967a), has found widespread practical application and is commonly known as the 'Huggett–Widdas formula'. The constant a in the formula is often referred to as the 'fetal growth constant'. If the cube root of fetal weight or some linear dimension of the fetus (W_d) is taken, Equation 5 can be converted to a linear form:

$$W_d = a^{0.33}.(t - t_0) \quad (6)$$

It is then a relatively simple matter to establish the values of a and t_o by fitting a regression line to the data (excluding early points that are likely to lie in the 'lag phase'). An example of this is provided in Fig. 6, illustrating growth of the placenta and the fetus in the common marmoset, *Callithrix jacchus*. Accordingly, where suitable data on fetal growth are available, empirical values of a and t_o can be directly determined for comparative purposes. It is also possible to derive a relationship between neonatal body weight (W_n) and gestation period (G) from the fetal growth formula. From Equation 5 it follows that, at the end of gestation, neonatal weight would be indicated by the following formula:

$$W_n = a.(G - t_0)^3 \quad (7)$$

Since t_o is a relatively small proportion of gestation period (estimated at approximately 20% overall by Payne & Wheeler 1967b, and Martin &

MacLarnon 1985), Equation 7 can be simplified for comparative purposes to:

$$W_n = a'.G^3 \tag{8}$$

where a' may be termed the approximated fetal growth factor.

As noted by Martin & MacLarnon (1985), there is an apparent conflict between Equation 4 and the result of combining Equations 2 and 8 to yield a general formula relating neonatal weight to maternal weight:

$$W_n = a'.(k.W_m{}^{0.10})^3 = a'.k^3.W_m{}^{0.30} \tag{9}$$

Equation 4 indicates a scaling exponent value of 0.85, whereas Equation 9 indicates an exponent value of only 0.30. This apparent paradox arises because the fetal growth constant a is only constant *within* a species and not *between* species of different body sizes, contrary to occasional assumptions in the literature (e.g. see Frazer & Huggett 1974). Values of a and a' are in fact found to vary in a systematic allometric fashion with maternal body weight (Rudder 1978, 1979; Martin & MacLarnon 1985). From Equations 4 and 9, it can be inferred that a' should be related to maternal body weight as follows:

$$a' \propto W_m{}^{0.55} \tag{10}$$

This theoretical expectation was confirmed through analyses of empirical data by Martin & MacLarnon (1985). It is therefore possible to explain the relationship between neonatal body weight (W_n) and maternal body weight (W_m) across placental mammals as a combined function of gestation time (G) and a fetal growth factor (a') reflecting the level of maternal investment in fetal growth. Both G and a' vary in a broadly predictable fashion with maternal body weight, though grade differences between groups of mammals must be taken into account. This observation can be summarized as follows:

$$W_n \propto \text{(maternal investment factor). (time factor)}$$

$$\begin{aligned} \text{hence:}\quad & W_n && \propto (a').(G^3) \\ \text{since:}\quad & a' && \propto W_m{}^{0.55} \\ \text{and:}\quad & G && \propto W_m{}^{0.10} \\ \text{therefore:}\quad & W_n && \propto W_m{}^{0.85} \end{aligned}$$

The relationships suggested can be further tested by analysis of the scaling of neonatal body weight to maternal body weight within groups of mammals in which gestation period is found to be relatively constant. This is illustrated, for instance, by intraspecific variation among domestic breeds of dogs and it is commonly found that gestation period varies little between

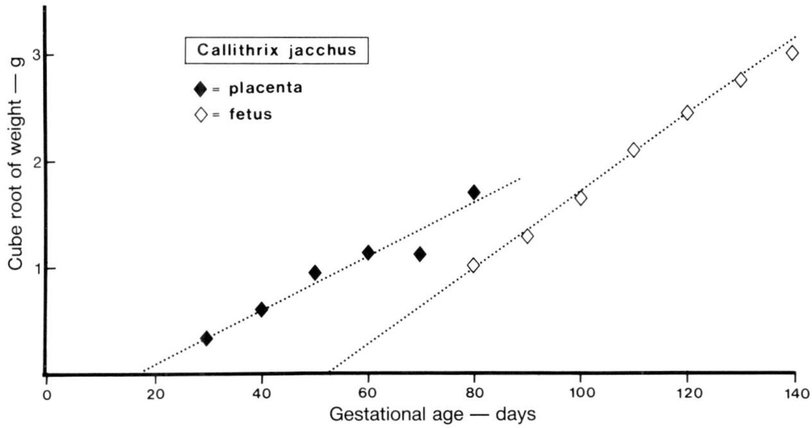

Fig. 6. Graph showing growth of placenta and fetus of the common marmoset (*Callithrix jacchus*) during gestation, taking cube roots of the weights. Dotted lines indicate regressions. The point of intersection of the fetal growth line with the abscissa indicates the value of the 'lag phase' (t_o) and the slope of the line indicates the value of the 'fetal growth constant' (a). (Original data from Chambers 1982.)

species within individual mammalian genera. If gestation period is held approximately constant across a group of mammals, the time factor disappears from the relationships described above and it can be predicted that the following relationship will be found between neonatal body weight (W_n) and maternal body weight (W_m):

$$W_n \propto W_m^{0.55} \tag{11}$$

This prediction was confirmed by Martin & MacLarnon (1985), who reported an exponent value of 0.56 for breeds of domestic dogs (see Kirkwood 1985) and figures yielding an average exponent value of 0.56 for five mammalian genera.

Having established this set of scaling relationships, it is possible to conduct broad comparisons of neonatal body weight and rate of maternal investment (both relative to maternal body weight) between mammalian species. In Fig. 7, distributions of residual values for neonatal body weight are illustrated for five orders of placental mammals. As expected (because of their markedly longer gestation periods), the two groups of precocial mammals (primates and artiodactyls) tend to have larger neonates, relative to maternal weight, than the altricial insectivores and carnivores. However, the difference is by no means as striking as for gestation periods and certainly does not reach the theoretically expected level of a 64-fold difference between typical precocial and typical altricial mammals. Further, there is a considerable degree of overlap in residual values between primates and

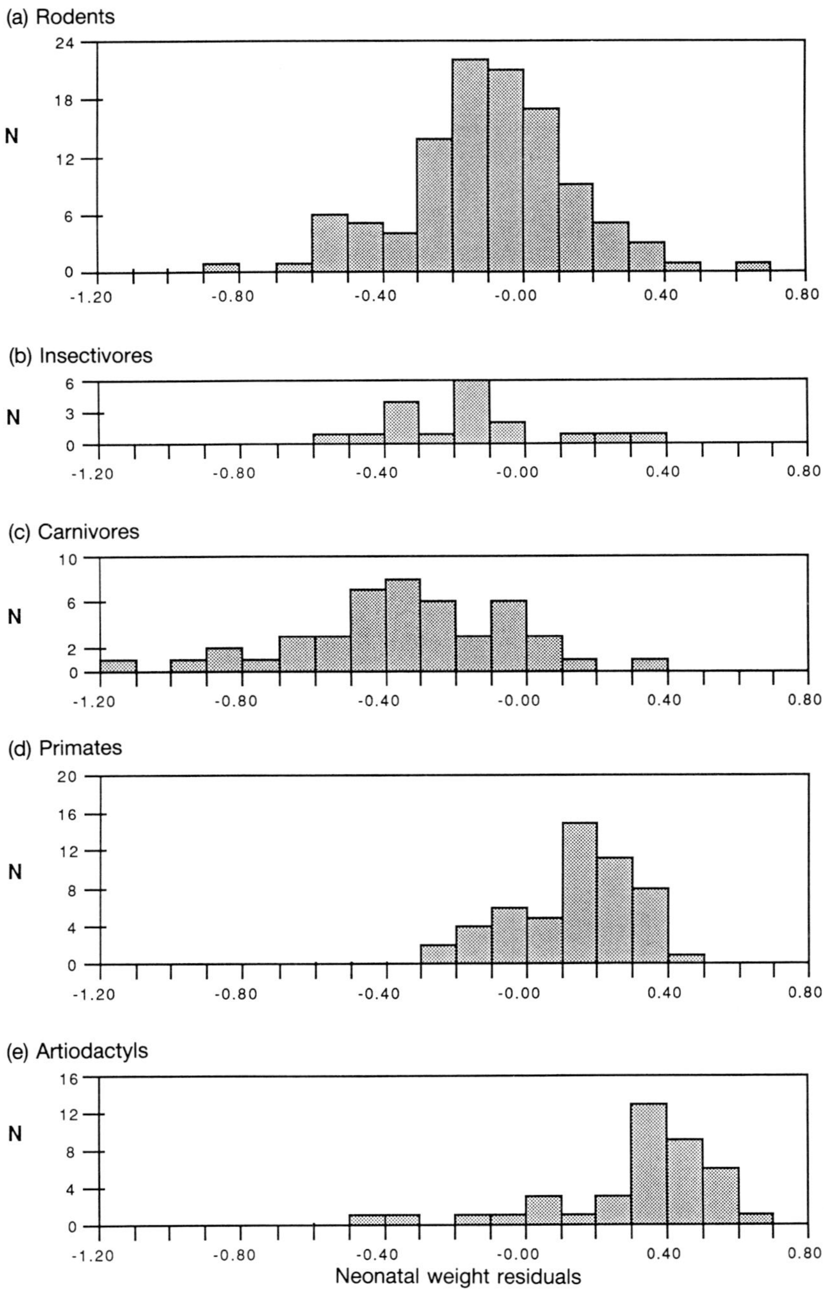

Fig. 7. Histograms of residual values for neonatal body weights, determined relative to a line of fixed slope 0.85, for five orders of placental mammals (Rodentia, Insectivora, Carnivora, Primates and Artiodactyla).

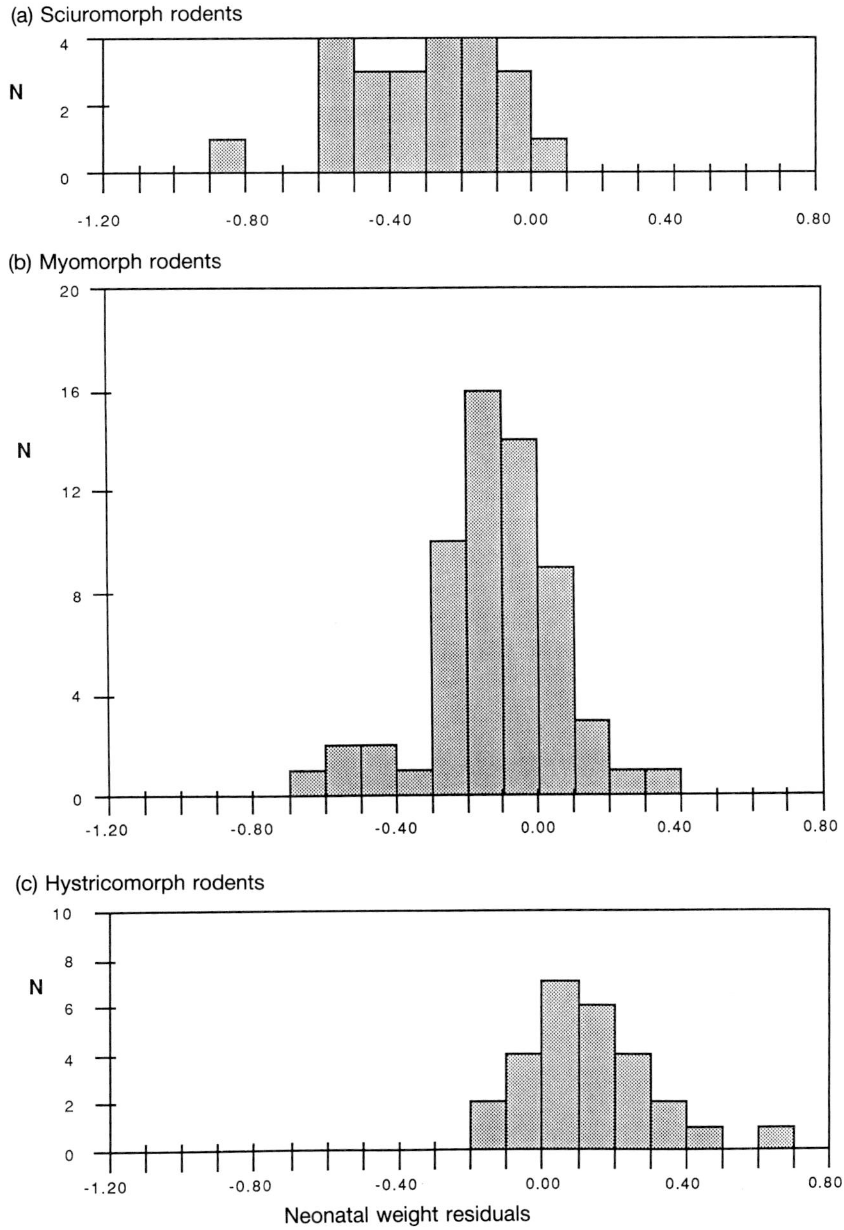

Fig. 8. Histograms of residual values for neonatal body weights, determined relative to a line of fixed slope 0.85, for the three suborders of Rodentia (Sciuromorpha, Myomorpha and Hystricomorpha). The typically altricial sciuromorphs and myomorphs show a pattern of distribution closer to that of insectivores and carnivores, while the typically precocial hystricomorphs show a pattern similar to that of artiodactyls.

insectivores or carnivores, while the artiodactyls generally have higher residual values than primates. This indicates that, in addition to differential effects of gestation period relative to maternal body size, there are differences between groups of mammals in the degree to which mothers allocate resources to fetal development per unit time.

Once again, the rodents show a broad distribution that overlaps both with typically altricial mammals (insectivores and carnivores) and with typically precocial mammals (primates and artiodactyls). Separate analysis of the three rodent suborders (Fig. 8) reveals that, as expected from their longer gestation periods relative to maternal size, hystricomorph rodents produce relatively larger neonates than myomorph or sciuromorph rodents. However, it is also noteworthy that the myomorph rodents, which have approximately the same distribution of relative gestation lengths as sciuromorph rodents, tend to produce relatively larger neonates. Indeed, the myomorphs overlap extensively with the hystricomorphs in Fig. 8, whereas there is only limited overlap between sciuromorphs and hystricomorphs. This indicates a difference between myomorph and sciuromorph rodents in the relative rate of allocation of maternal resources during gestation.

Patterns of neonatal weight residuals therefore raise the issue of differential maternal investment. However, in order to reach a satisfactory comparative assessment of maternal investment, it is necessary to take account of the effects of maternal body weight and gestation period simultaneously. In other words, it is necessary to devise some index of differences between mammalian species with both gestation period and maternal size held constant. One approach to this, suggested by Martin & MacLarnon (1985) is to examine values of a' (the approximated fetal growth factor), which can be derived by rearranging Equation 8 as follows:

$$a' = W_n \div G^3 \qquad (12)$$

However, the relationship between a' and neonatal body weight (W_n) is complicated by the fact that the resources supplied by the mother must increase in an approximately cubic fashion throughout gestation. Hence, a' refers to a basic 'setting' of the fetal growth curve for any individual species but does not give a clear indication of the average cost to the mother throughout gestation. An alternative approach, suggested by Payne & Wheeler (1967b), is simply to divide neonatal body weight by the gestation period. This provides an estimate of maternal resources (R) devoted to fetal growth per day during gestation. This measure R is given by the following simple equation:

$$R = W_n \div G \qquad (13)$$

From Equations 2 and 4 above, it follows that R should theoretically be related to maternal body weight in the following way:

$$R \propto W_m^{0.85} \div W_m^{0.10} \tag{14}$$

viz.

$$R = k.W_m^{0.75} \tag{15}$$

The derived relationship in Equation 15 is of particular interest since it is widely accepted that basal metabolic rate (=standard metabolic rate) scales to body weight with an exponent value close to 0.75 (Benedict 1938; Brody 1945; Kleiber 1947, 1961). This regular pattern of scaling of basal metabolic rate to body weight in mammals is commonly referred to as 'Kleiber's Law'. Hence, it may be inferred that there is a link between basal metabolic rate and maternal resources devoted to fetal growth per day of gestation. Such a link is only to be expected, as total maternal resources are likely to scale to maternal body weight with an exponent value close to 0.75 and as, other things being equal, one might expect mammalian mothers to devote a standard proportion of their resources to cover the costs of fetal growth. In addition, the capacity of a mother to supply resources for fetal growth during gestation will clearly be related to her general metabolic turnover, not only when she is active but also during inactivity (for which basal metabolic rate is relevant).

On the basis of Equation 15 and its agreement with expectation from the scaling of basal metabolic rate, residual values for rate of maternal investment in individual fetal growth can be determined by fitting a line of fixed slope = 0.75 to data for neonatal body weight divided by gestation period ($W_n \div G$). The patterns of residual values obtained for five orders of placental mammals are illustrated in Fig. 9. From this Figure, it can be seen that carnivores and primates tend to have the lowest values, and artiodactyls, insectivores and rodents (in that order) tend to have the highest values. When the rodents are examined in more detail (Fig. 10), it emerges that the sciuromorphs and the hystricomorphs show broadly similar distributions of residual values, while the myomorphs typically show higher values. Indeed, the values for myomorphs include the highest shown for the five orders of placental mammals in Fig. 9. Hence, Fig. 10 indicates that the myomorph rodents are adapted for particularly high daily maternal investment in individual fetuses.

The analyses shown in Figs 9 and 10 refer only to maternal investment in the growth of the individual fetus and take no account of the average number of fetuses in the uterus (the minimum value of which is indicated by average litter-size). In order to estimate total daily maternal investment in fetal growth, it is necessary to multiply the value of R by litter-size (see also Payne & Wheeler 1967b). Since altricial mammals (e.g. carnivores, insectivores, myomorph rodents) typically have multiple litters and since precocial mammals (e.g. artiodactyls, primates) typically produce single neonates, it is clear that mothers of altricial offspring typically invest considerably more

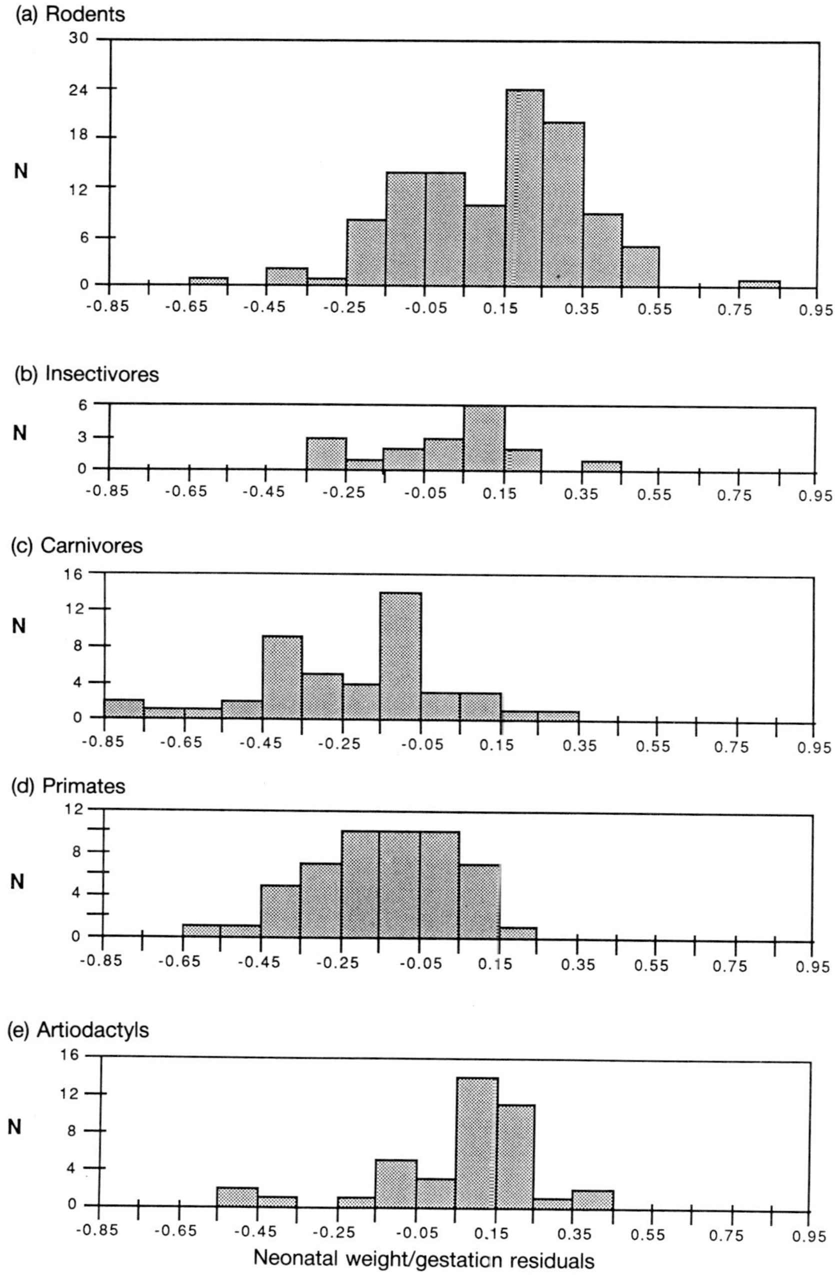

Fig. 9. Histograms of residual values for neonatal body weights divided by gestation periods, determined relative to a line of fixed slope 0.75, for five orders of placental mammals (Rodentia, Insectivora, Carnivora, Primates and Artiodactyla).

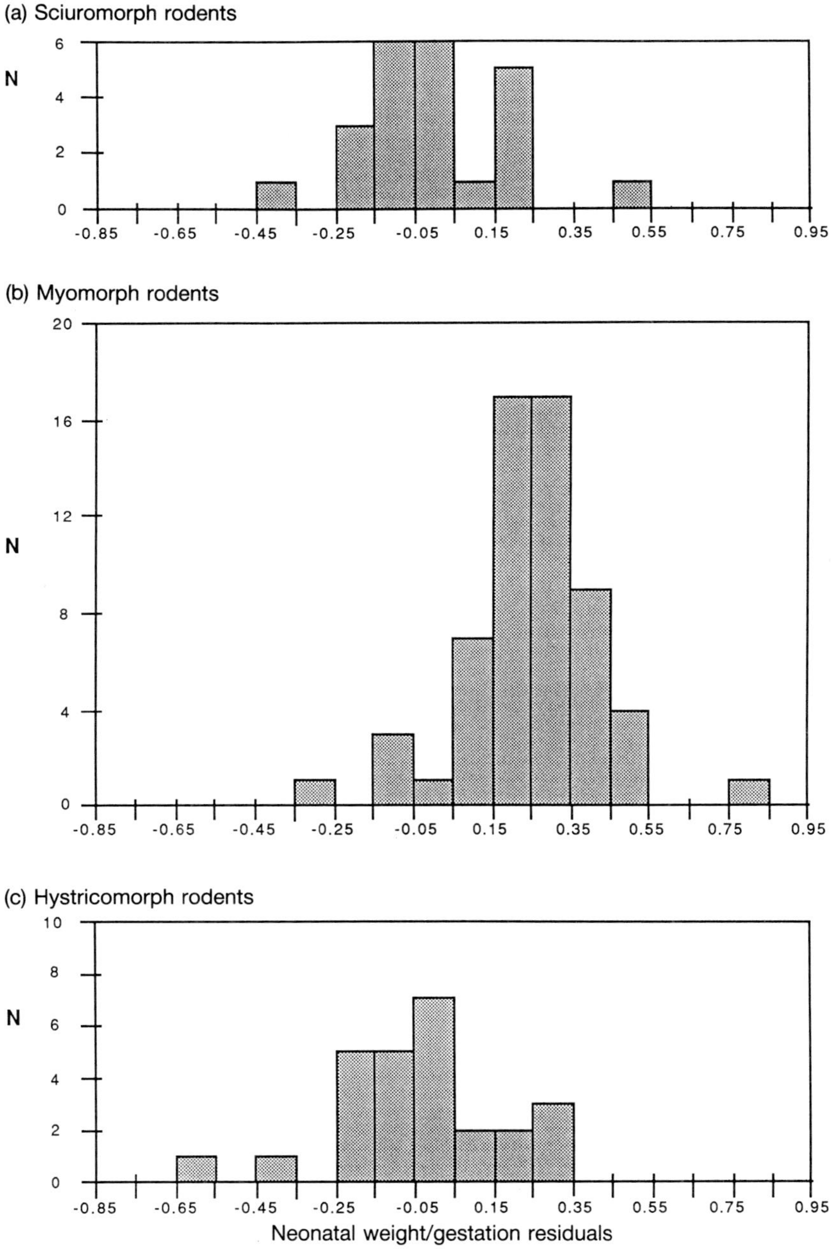

Fig. 10. Histograms of residual values for neonatal body weights divided by gestation periods, determined relative to a line of fixed slope 0.75, for the three suborders of Rodentia (Sciuromorpha, Myomorpha and Hystricomorpha). In this case, the altricial sciuromorphs resemble the precocial hystricomorphs in their distribution, while the myomorphs show generally higher values for this measure of rate of maternal investment in individual offspring.

resources in fetal growth per day of gestation (relative to maternal body weight) than mothers of precocial offspring. This conclusion is in agreement with that reached by Martin & MacLarnon (1985) from an analysis of values of a' for altricial and precocial mammals. Further, once litter-size has been taken into account along with the measure of maternal investment in the individual fetus (R), it emerges that primates exhibit very low levels of maternal investment among mammals generally. This, in fact, is matched by very low relative postnatal levels of investment in lactation (Martin 1984a) and the overall picture is that primate mothers allocate a relatively small proportion of their total daily resources to reproduction (see also Ross 1988).

Some comment is needed on the meaning of the term 'maternal investment' as used here (see also Martin & MacLarnon 1986). The assumption implicit in the discussion above is that maternal investment should be seen in terms of differential allocation of resources per unit time. In other words, it is the proportion of total maternal resources allocated to reproduction, rather than to other functions, that is at stake. The measure R given in Equation 13 provides one simple measure of the resources devoted per day to the growth of the individual fetus throughout gestation and multiplication of this by litter-size gives an approximate indication of the total maternal resources devoted to fetal growth per day during gestation. Zeveloff & Boyce (1986) have questioned this approach to maternal investment and have pointed out, correctly, that total investment in individual young is greater for precocial offspring than for altricial offspring. If one is concerned with total maternal investment in individual young by the time of birth, regardless of the time taken to make that investment, then precocial mammals undoubtedly surpass altricial mammals. However, if one is concerned with the *rate* of maternal investment in reproduction and, particularly, with the differential allocation of resources to reproduction, then altricial mammals undoubtedly surpass precocial mammals (see also Ross 1988). Since the proportion of resources devoted to reproduction, rather than total investment in individual offspring, is more generally relevant in terms of ecological energetics, 'maternal investment' is used in this sense here. It should be noted, incidentally, that Zeveloff & Boyce (1980, 1986) *divided* neonatal weight by litter-size to obtain a measure of 'maternal investment'. Although they justify this on the grounds that the measure reflects the degree of competition between siblings, it is difficult to see the practical value of such a measure. Within a species, for example, a mother producing six infants weighing 12 g each (a total neonatal mass of 72 g) would be said to have made an 'investment' only one-fifth as large as that made by a mother of the same body size producing two infants of 20 g each (a total neonatal mass of 40 g). In such a case, where all other major factors are maintained constant, it is surely obvious by *any* definition that the mother with the more numerous litter has invested more resources in reproduction!

The question of maternal investment in relation to neonatal body weight is also specifically relevant to an interesting finding in relation to primates made by Leutenegger (1973). It was found that strepsirhine primates (lemurs and lorises) typically have much smaller neonates, relative to maternal body weight, then haplorhine primates (tarsiers, monkeys, apes and man). There is a clear separation between the two grades (strepsirhines *versus* haplorhines) in a bivariate allometric plot and the difference corresponds to an almost threefold difference in neonatal body weight (for a given maternal weight) between haplorhines and strepsirhines (see also Martin 1975). This difference cannot be attributed to a systematic difference in gestation period, relative to maternal body size, between strepsirhines and haplorhines. Although some strepsirhine primates do have relatively short gestation periods (e.g. *Cheirogaleus, Microcebus, Varecia*), most overlap with the distribution for haplorhine primates (Martin 1975). However, strepsirhines do differ significantly from haplorhines in their placentation; whereas lemurs and lorises have a non-invasive, epitheliochorial type of placentation, haplorhine primates have a highly invasive, haemochorial type. Leutenegger (1973) accordingly suggested a direct link between placental type and relative neonatal size in primates. He proposed that the more invasive haemochorial placentation of the haplorhine primates permits more rapid transfer of nutrients from mother to fetus and hence production of a larger neonate.

The difference between strepsirhine primates (lemurs and lorises) and haplorhine primates (tarsiers, monkeys, apes and man), originally identified by Leutenegger (1973) in a study devoted to primates alone, is still clearly apparent in the residual values derived from an analysis of the entire sample of placental mammals (Fig. 11), though there is some overlap between the two distributions. Further, for primates generally, there is only a very weak and not quite significant correlation between residuals for neonatal weight and residuals for gestation ($r = 0.26$; $r^2 = 0.07$; $N = 52$; $p < 0.06$). Hence, overall the relatively larger neonatal size of haplorhine primates cannot be attributed to relatively longer gestation periods. At first sight, therefore, it might seem that Leutenegger's hypothesis is confirmed. However, it must be remembered that the apparent link between small relative neonatal size and non-invasive placentation in strepsirhine primates is a *correlation* and that a causal link remains to be demonstrated. It is, for instance, still necessary to take into account the influence of gestation period, since some strepsirhines (*Cheirogaleus, Microcebus, Varecia*) have much shorter relative gestation lengths than haplorhine primates. The effect of gestation period can be excluded by examining values of the ratio of neonatal body weight to gestation period (Fig. 12). When this is done, although it remains true (as

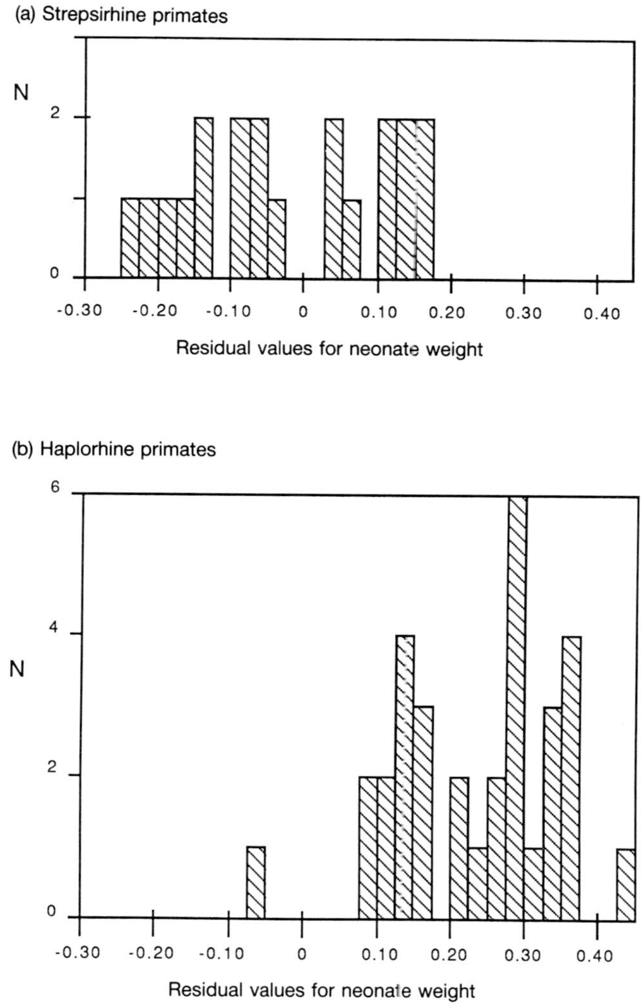

Fig. 11. Histograms of residual values for neonatal body weights, determined relative to a line of fixed slope 0.85, for the two main subgroups of living primates, strepsirhines and haplorhines. Strepsirhines typically have smaller neonates, relative to maternal body weight, than haplorhines and there is only limited overlap between the two distributions. (The aberrant haplorhine with a residual value of less than zero is, in fact, the gorilla.)

expected) that most strepsirhine primates have lower values than haplorhine primates, there is more overlap than for neonatal body weights alone (Fig. 11). Further, two strepsirhine primates (*Microcebus murinus* and *Cheirogaleus medius*) are found to have very high values for the ratio between neonatal weight and gestation period. Indeed, *Cheirogaleus medius* is found to have the highest value of all primates examined. Hence, it is clear

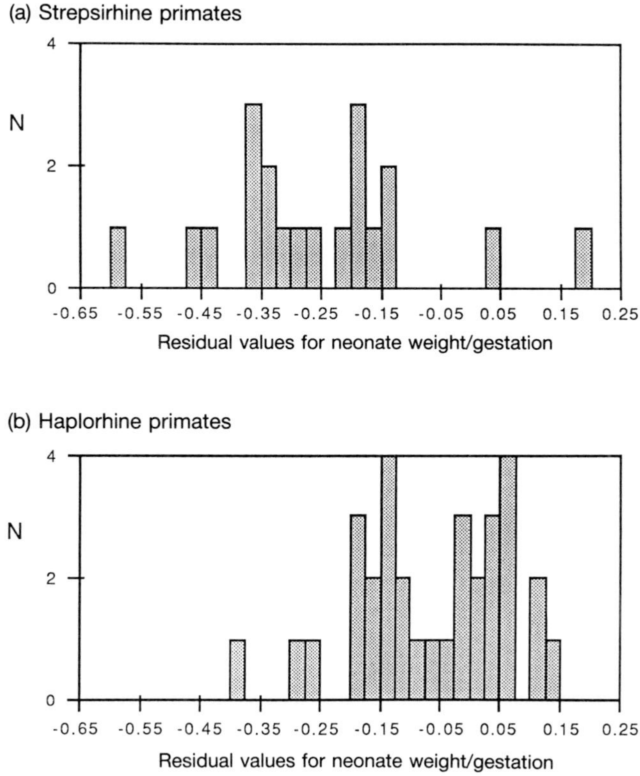

Fig. 12. Histograms of residual values for neonatal body weights divided by gestation periods, determined relative to a line of fixed slope 0.75, for the two main subgroups of living primates, strepsirhines and haplorhines. Although, as with neonatal body weights, strepsirhines typically have smaller values for the ratio between neonatal weight and gestation, some strepsirhines have notably high values. Indeed, the highest value recorded is for a strepsirhine (*Cheirogaleus medius*), not for a haplorhine.

that *Microcebus* and *Cheirogaleus*, despite the non-invasive nature of their placentation, are able to generate fetal tissue as effectively as haplorhine primates per unit time. This suggests that it is not the structure of the placentation but some other factor that limits the rate of fetal growth in most strepsirhine primates.

Inclusion of other placental mammal groups in the analysis in fact permits further testing of Leutenegger's (1973) hypothesis, since there is a clear prediction that non-invasive placentation among placental mammals generally should be associated with small neonatal sizes (once the effects of gestation period and maternal body size have been taken into account). As has already been noted by Martin (1984a), precocial mammals generally tend to have larger neonates than altricial mammals (relative to maternal body size), because of their longer gestation periods. However, when precocial mammals alone are taken, it is found that artiodactyls, perissodactyls, pinnipeds and cetaceans (whales and dolphins) all tend to produce larger neonates than *any* primates, despite the fact that there is comparatively little difference in gestation periods relative to maternal size (e.g. see Fig. 4 for primates and artiodactyls). Figure 7, for instance, shows that many artiodactyls have larger residual values for neonatal body weight than any primate, including haplorhines. This finding is, in fact, slightly enhanced when residual values for the ratio of neonatal weight to gestation period are taken (Fig. 9). Yet artiodactyls (in common with perissodactyls and cetaceans) have non-invasive epitheliochorial placentation. Hoofed mammals (artiodactyls and perissodactyls), cetaceans and pinnipeds produce the largest neonates among placental mammals, both relatively and absolutely. Hence, for placental mammals generally there is no evidence that the non-invasive epitheliochorial type of placentation constrains either the rate or the extent of fetal development.

It is a common assumption in the literature that the degree of invasiveness of the placenta (along the continuum passing from epitheliochorial through endotheliochorial to haemochorial) is associated with differential efficiency of transfer of materials between the mother and fetus (e.g. see Payne & Wheeler 1967b). This is, indeed, one of the reasons why epitheliochorial placentation, as found in modern mammals, is generally held to be more primitive than the invasive forms of placentation (endotheliochorial; haemochorial). However, there is little direct evidence to indicate that the type of placentation constrains the rate of development of the fetus in the uterus and the comparative evidence presented above indicates that this is not, in fact, the case. An alternative view is that the overall rate of fetal development is so slow that the rate of transfer of materials across the placenta is not usually a major limiting factor. According to this view, it is more fruitful to consider differences in rate of fetal development between species in terms of differential allocation of maternal resources. In the primates, for instance, explanation of the fact that strepsirhine mothers typically produce smaller neonates than haplorhine mothers should be sought in terms of different patterns of resource allocation. In other words, a major factor determining the relative size of the neonate is the differential allocation of resources by mothers of different species. The possibility that such differential allocation of resources occurs is supported by analysis of the relationship between neonatal weight residuals and gestation residuals in strepsirhines and haplorhines separately. In neither case is there a significant correlation (stepsirhines: $r = 0.05$, $N = 20$, $p = 0.83$; haplorhines: $r = 0.15$, $N = 32$; $p = 0.59$). Since the type of placentation is approximately constant within each group of primates, variations in relative neonatal weight between species must be explained by some factor other than placentation or gestation period. That factor could either be a direct metabolic influence that constrains the availability of resources or an indirect ecological influence of some kind that ultimately determines the rate of maternal investment. It has, for example, been suggested that strepsirhine primates exhibit certain features indicative of adaptation to marked seasonal changes at some time during their evolutionary history (Martin 1975). This could account for an overall pattern in which the commitment of relatively limited energy to fetal growth is compensated by relatively rapid postnatal growth.

The influence of metabolic rate

It has already been noted above that there is a widely recognized scaling relationship between basal metabolic rate and adult body weight, with an exponent value close to 0.75 ('Kleiber's Law'). It has also been shown that the scaling of maternal investment during gestation, as indicated by the ratio of neonatal body weight to gestation period, should theoretically exhibit a very similar exponent value. For this reason, it was suggested that, other things being equal, there might be a direct link between the scaling of maternal resources generally and the scaling of resources specifically devoted to fetal development during gestation. Once again, however, the correspondence in scaling between basal metabolic rate and the measure of maternal investment is no more than a correlation and further investigation is required to confirm or deny the presence of a causal link.

Although there is a fairly regular relationship between basal metabolic rate and body weight for mammals generally, as expressed in 'Kleiber's Law', it is possible to recognize grade effects and individual outliers. For instance, in marsupials basal metabolic rate scales to body weight with an exponent of 0.74, but marsupials as a group represent a lower grade in comparison with placental mammals (Dawson & Hulbert 1970). In practical terms, a typical marsupial will consume 30% less energy per unit time for basal metabolism than a typical placental mammal of the same body weight (see Martin 1984b). There are also individual outliers among the placental mammals, the best known example being that of the sloths, which have basal metabolic rates less than 50% of the values predicted by the Kleiber equation and hence have lower basal metabolic rates than typical marsupials (McNab 1978). McNab (1980, 1986) has suggested that relatively low metabolic rates among placental mammals can be linked to specific food habits, such as folivory, though Elgar & Harvey (1987) have shown that taxonomic associations may account for much of the variation in basal metabolic rate. In fact, after controlling for taxonomic effects, they found that the only consistent relationship was between consumption of vertebrate prey and relatively high metabolic rates. Among the primates, there are also a number of well-known examples of relatively low basal metabolic rates, such as the potto (*Perodicticus*), the slow loris (*Nycticebus*) and the owl monkey (*Aotus*). In fact, in a recent review, Müller (1983) has shown a taxonomic effect among primates in that strepsirhine primates generally show lower basal metabolic rates, relative to body size, than simian primates (monkeys, apes and man). Whereas simian primates exhibit basal metabolic rates close to those predicted by the Kleiber equation, strepsirhine primates exhibit values that are at least 20% less than predicted. The potto and the slow loris are especially divergent in this respect. In a subsequent review, Müller (1985) showed that the lower basal metabolic rates of strepsirhine primates are probably not primitive, as has commonly been assumed, but represent special adaptations to particular ecological conditions. More recently, Richard & Nicoll (1987) have reported on the resting metabolic rate of the Malagasy lemur *Propithecus verreauxi*, showing that this species conforms with other strepsirhine primates in exhibiting a low residual value. These authors also link low basal metabolic rate to ecological factors.

If there is a link between the scaling of basal metabolic rate and the scaling of maternal investment in fetal growth, it can be predicted that species with a relatively low basal metabolic rate should exhibit relatively low rates of maternal investment during gestation. This can be tested by determining residual values for basal metabolic rate (relative to a line of fixed slope = 0.75) and by subsequently examining the relationship between these values and residual values for the ratio of neonate weight to gestation period. After taking available data on basal metabolic rate for 18 primate species and calculating the residual values, it was found that there is indeed a fairly consistent relationship with residual values derived from the ratio between neonatal weight and gestation period, also taken relative to a line of fixed slope = 0.75 (Fig. 13). The correlation coefficient for this relationship is $r = 0.72$, which is highly significant ($p < 0.001$). This indicates that variations in basal metabolic rate, relative to maternal body weight, account for more than 50% of the variation in rate of maternal investment in fetal growth (coefficient of determination = 0.52). There is, therefore, indirect evidence that maternal metabolic rate constrains the rate of fetal development. Further, the fact that strepsirhine primates (lemurs and lorises) consistently show lower basal metabolic rates than simian primates (monkeys, apes and man) may therefore be linked to the observation (Fig. 12) that strepsirhines show relatively lower rates of maternal investment than haplorhines (tarsiers + simians). Hence, the difference between strepsirhines and haplorhines in the size of the neonate relative to maternal body size (Fig. 11) might be ascribed at least in part to the lower basal metabolic rates of strepsirhine primates. Past or present ecological factors influencing rates of maternal investment probably account for the rest of the difference between strepsirhines and haplorhines, whereas the type of placentation would not seem to exert any measurable effect.

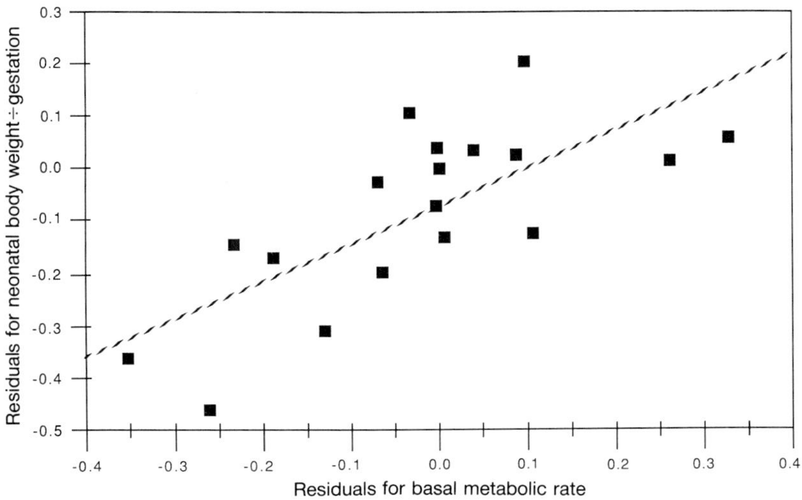

Fig. 13. Plot of residual values for neonatal body weights divided by gestation periods against residual values for basal metabolic rate, both determined relative to a line of fixed slope 0.75, for a sample of 18 primate species. The regression line (dashed) indicates the general trend.

Discussion

The fundamental tenet of this paper is that the influence of body size must be taken into account for any meaningful comparison of biological parameters between species of markedly different body sizes. This is true both in a general biological framework and within a specific medical context in which, for example, extrapolations may be made from small-bodied laboratory rodents to large-bodied human subjects.

It is, of course, true that uncertainty continues to reign over the most appropriate techniques for identifying and 'removing' the influence of body size in comparative studies. The basic method advocated here is that of allometric analysis and for this method there are two major remaining problems of a statistical nature. In the first place, it is necessary to decide upon the correct choice of a procedure for determining a best-fit line (e.g. regression *versus* major axis). There are, as yet, no unequivocal guidelines for making such a decision. Secondly, there is the problem that a given data set may include distinct subsets ('grades') and that determination of a *single* best-fit line may not be appropriate. Both problems may in some instances be avoided by fitting a line of *fixed* slope to the data on the basis of prior theoretical expectations. A possible additional solution to the second problem is suggested here, based on iteration of slope (exponent) values and examination of the patterns of residuals generated. This latter approach has yielded very encouraging results in some cases (e.g. with the scaling of gestation periods), but it may not always produce a clear solution (e.g. with the scaling of neonatal body weights).

Whatever problems are encountered in identifying an appropriate technique for eliminating body size effects in interspecific comparisons, there is certainly no justification for making such comparisons *without* any attempt to take scaling into account. Numerous mistakes have been made in the literature because individual parameters (e.g. gestation period) have been compared between species without taking into account the fact that, other things being equal, large-bodied species must be expected to have larger values than small-bodied species. When body size effects are taken into account in the analysis of gestation periods (Figs 2 and 3), it can be shown that the placental mammals fall clearly into two groups, corresponding fairly well to the 'altricial' and 'precocial' categories recognized by Portmann (1938, 1939, 1965). There is, therefore, a basic dichotomy between mammals that produce large litters of poorly developed offspring after a relatively short gestation period and mammals that produce small litters of well developed young after a relatively long gestation period. Primates, including man, belong to the latter category. Although human infants are often referred to as 'altricial', they are clearly classified in the precocial category here on the grounds of typical single births and opening of the eyes by the time of birth. The relative helplessness of human babies at birth, as compared to infants of other primate species, can be attributed to a special pattern of brain development in which fetal growth relationships are continued for approximately a year after birth (Portmann 1941; Martin 1983). It is therefore important to note that some laboratory rodents commonly used in medical research (e.g. mice, rats) are *extreme* representatives of the altricial category and that a newborn rat is not directly comparable with a human neonate. Considerable care is therefore required in the extrapolation of observations on fetal and postnatal development in most laboratory rodents to human biology. In fact, the evidence indicates that hystricomorph rodents (e.g. the guinea-pig) should be more directly comparable to primates than myomorph rodents, though even here scaling relationships must be respected.

In fact, because the distinction between 'altricial' and 'precocial' mammals with respect to gestation period relative to maternal body size is so great (an approximately fourfold difference) and because the scaling exponent determined for gestation period (0.10) is so low, there is a large degree of separation between the two groups even without taking body size into account. However, if a histogram of logarithmic values for gestation periods is compiled (i.e. effectively taking a zero exponent value for scaling), a *trimodal* distribution is obtained. This is because large-bodied altricial mammals have gestation periods similar to those of small-bodied precocial mammals and therefore produce an intervening peak that disappears when the scaling effect of body size is taken into account (Martin & MacLarnon 1986). On the one hand, the large degree of separation between altricial and precocial mammals found even without taking body size into account underlines the validity of this distinction. On the other hand, the confusion of large-bodied altricial mammals with small-bodied precocial mammals that occurs when body size is not taken into account provides a good illustration of the dangers of ignoring scaling effects.

Through interspecific analyses of the scaling of gestation periods and neonatal body weights, a number of general relationships have been established. These may be used for more effective comparison between species and they may also be applied in a wider context. In particular, combination of the results from the *interspecific* comparisons with information derived from intraspecific studies of fetal growth permits the formulation of a general equation that described fetal (and, ultimately, neonatal) body size in terms of a time component and a second component representing the relative allocation of maternal resources. In order to develop a comparative measure of rate of 'maternal investment', in terms of commitment of resources per unit time, it is necessary to take into account both maternal body size and time elapsed.

One simple relative measure of rate of maternal investment is provided by the ratio of neonate weight to gestation period. This provides a crude indication of the maternal resources devoted to fetal growth per day. Of course, maternal resources are also required for accessory structures associated with fetal development (e.g. placenta, fetal membranes), but these seem to scale approximately in direct proportion to fetal body weight, so that the latter may be taken as a reliable indicator of the overall investment in fetal growth (Payne & Wheeler 1967b). On the basis of equations derived for the scaling of gestation period and neonatal body weight to maternal body size, it is shown above that the ratio of individual neonate size to gestation period may theoretically be expected to scale to maternal body weight with an exponent value of 0.75. This value coincides with that applying to the scaling of basal metabolic rate to body weight in mammals ('Kleiber's Law') and there is therefore a possibility that this measure of maternal investment may be directly linked to metabolic capacity of the mother. Evidence is, in fact, provided to confirm this link for primates, in that the residual values for this measure of maternal investment are quite tightly correlated with residual values for basal metabolic rate.

It should be noted, however, that Payne & Wheeler (1967b) did not find the expected exponent value of 0.75 when they scaled the ratio of total neonate weight to gestation period against maternal body weight for a small sample of placental mammals. In agreement with the findings reported here, they found that primates show a lower level of investment than other placental mammals. However, when they fitted regression lines to primates and to other mammals separately, they obtained equal slope values of 0.54. This common value was explained in terms of an interaction between metabolic scaling (exponent = 0.75) and surface scaling of the placenta (exponent = 0.67), giving a combined exponent value of approximately 0.50. It should be noted, however, that the data analysed by Payne & Wheeler (1967b) were for total mass of the litter and not for individual neonatal weight. The line fitted to 'other mammals' combined altricial mammals (which tend to be small) with precocial mammals (which tend to be big). Since it has been shown above that altricial mammals show a higher rate of maternal investment in individual neonates, relative to maternal body weight, than precocial mammals, and also have multiple litters, it is only to be expected that a line fitted to both groups together would have a lower slope than lines fitted to altricial and precocial mammals separately. This provides another example of 'grade confusion' (see Fig. 1b). The low slope value of 0.54 found for the primate sample by Payne & Wheeler (1967b) can be attributed to a number of factors. In the first place, their sample size was very small (only 16 species, as compared to the 52 included in the present study) and the data were of a very preliminary nature. One particularly significant error was that the neonatal body weight of *Lemur catta* was reported to be 240 g, whereas the substantial data now available indicate a mean neonatal body weight of only 86 g. Further, the common tree-shrew, *Tupaia belangeri* (= '*Tupaia glis*') was included among the primates, although it has typically altricial offspring in contrast to the primates. In fact, the residual value determined for the ratio of neonatal weight to gestation period for *Tupaia belangeri* was 0.11. This value falls well within the range for insectivores and lies at the upper end of the distribution for primates (see Fig. 9). The combined effect of including tree-shrews in the primate sample and taking a neonatal body weight for *Lemur catta* that was three times too large was to lower the slope of the best-fit line, since both species are relatively small in body size and the values used acted to 'tip' the line.

The calculation by Payne & Wheeler (1967b) of the expected exponent value of 0.50 for the scaling of the ratio of neonatal weight to gestation period was directly linked to the common assumption that the placenta limits the rate of fetal growth. In their case, it was assumed that the surface area of the placenta would influence fetal growth and many authors have assumed that the type of placenta would also be a determinant of fetal growth rate. It has been widely assumed that non-invasive, epitheliochorial placentae are more primitive and less efficient than invasive, haemochorial placentae and that rates of diffusion across the latter are much greater. Although it may seem obvious that invasive placentae permit faster rates of exchange between mother and fetus, there is little experimental evidence to support this assumption. Further, even if true, this would affect rates of fetal growth only if rate of diffusion is the limiting factor. If some other factor—such as the metabolic turnover of the mother or selection for the level of resources allocated to production—limits the rate of fetal growth, the type of placentation may be largely irrelevant to the rate of fetal growth. The idea that the type of placentation influences the rate of fetal growth was apparently supported by Leutenegger's (1973) report that strepsirhine primates, which have non-invasive placentation, typically produce considerably smaller neonates than haplorhine primates, which have invasive placentation. However, this correlation does not hold up when other placental mammals are examined. In fact, the largest neonates (both relatively and absolutely) found among placental mammals are produced by cetacean and ungulate species with non-invasive, epitheliochorial placentation. Various other lines of evidence indicate that it is not the type of placenta that limits fetal growth but a combination of ecological factors and the relative setting of maternal metabolic turnover, which together determine the rate of maternal investment in fetal growth.

Evidence has been presented above to show that the rate of maternal investment in fetal growth scales in the same way as basal metabolic rate and that *relative* levels are found to be correlated (in primates, at least) when the effect of body size has been removed. There are, however, at least two problems associated with the use of basal metabolic rate as an indicator of maternal metabolic turnover. Ideally, the allocation of maternal resources should be viewed in relation to data on total metabolic turnover of the mother (including the costs of activity), rather than merely to data on basal metabolic turnover. There is some evidence that active metabolic rate may not necessarily scale to body size in the same way as basal metabolic rate (see Martin 1984b). However, the relationship between energetic costs per day for activity and total metabolic turnover is at present unclear. Further, reliable data on active metabolic rate or on total metabolic turnover are considerably scarcer than reliable data on basal metabolic rate, which are in any case far from plentiful. For the time being, therefore, it is only possible to examine the relationship between basal metabolic rate and reproductive parameters, in the hope that total metabolic turnover scales in a fashion broadly similar to that of basal metabolic rate. The second problem is that basal metabolic rate may be altered during pregnancy. There is evidence to show that, in some mammalian species at least, a basal metabolic rate that is otherwise below the level predicted by the Kleiber equation will be raised approximately to that level during pregnancy (see Thompson & Nicoll 1986; Richard & Nicoll 1987). Thus, species that generally show a relatively low basal metabolic rate may not exhibit the expected relatively low rate of maternal investment in fetal growth. This might explain some of the deviations found in the plot of residuals for maternal investment rate against residuals for basal metabolic rate in primates (Fig. 13). In particular, it is striking that the lesser mouse lemur (*Microcebus*), the dwarf lemur (*Cheirogaleus*) and the owl monkey (*Aotus*) all show relatively high ratios of neonate weight to gestation period (among the top 12 out of 52 primate species), despite the relatively low basal metabolic rates measured for non-pregnant females. It can therefore be predicted on the basis of evidence presented in this paper that all three genera exhibit significant raising of basal metabolic rates during pregnancy. Conversely, the very low ratio of neonatal weight to gestation period found in the gorilla indicates that basal metabolic rates may be low in this primate species and that there is no increase during pregnancy. These cases illustrate the important point that empirical allometrical analyses may yield specific predictions that can be tested by detailed studies of single species.

As a final point it should be noted that confusion may arise in referring to 'constraints' or 'limitations' on fetal growth imposed by maternal metabolic turnover. On the one hand, it is possible that a direct constraint is imposed because a mother from a species with a relatively low basal metabolic rate would automatically be subject to decelerated fetal growth in the absence of special mechanisms (e.g. increase in basal metabolic rate during pregnancy). With such direct constraints, one would expect a very close relationship between metabolic scaling and scaling of maternal investment rate, though species with special adaptations would be outliers. However, there is an indirect sense in which metabolic turnover may impose constraints. Assuming that metabolic turnover provides a good indicator of the total resources available to the mother, different patterns of partition of metabolic resources found in different species may represent long-term evolutionary change, in response to specific environmental conditions. In such cases, one might expect that mammals would show a general scaling of maternal investment rate in accordance with Kleiber's Law (viz. with an exponent value of 0.75) but with numerous outliers and grades. It is clear from the evidence above that both direct and indirect factors seem to play a part in the determination of maternal investment rate among placental mammals, though it is possible that direct factors are more important for the individual fetus and that indirect factors are more likely to influence parameters such as litter-size.

Acknowledgements

The research on which this paper is based was supported by the Medical Research Council. For help in providing data or material we thank the Jersey Wildlife Preservation Trust (Jersey Zoo), East Midlands Zoological Society (Twycross Zoo), and The Zoological Society of London. For helpful discussion we thank Dr Leslie Aiello, Mr Fred Brett, Dr Paul Harvey, Dr Michael Hills, Dr Georgina Mace, Dr Phillip Payne, Dr Ben Rudder and Dr Erika Wheeler.

References

Benedict, F.G. (1938). Vital energetics: a study in comparative basal metabolism. *Publs Carnegie Instn* **503:** 1–215.

Blueweiss, L., Fox, H., Kudzma, V., Nakashima, D., Peters, R. & Sams, S. (1978). Relationships between body size and some life history parameters. *Oecologia* **37:** 257–72.

Brody, S. (1945). *Bioenergetics and growth*. Reinhold Pub. Co., New York.

Calder, W.A. (1982). The pace of growth: an allometric approach to comparative embryonic and post-embryonic growth. *J. Zool., Lond.* **198:** 215–25.

Calder, W.A. (1984). *Size, function and life history*. Harvard University Press, Cambridge, Mass. & London.

Chambers, P.L. (1982). *The endocrinology of pregnancy in the marmoset monkey,* Callithrix jacchus. PhD Thesis: University of Edinburgh.

Dawson, T.J. & Hulbert, A.J. (1970). Standard metabolism, body temperature, and surface areas of Australian marsupials. *Am. J. Physiol.* **218:** 2133–8.

Dorst, J. & Dandelot, P. (1970). *A field guide to the larger mammals of Africa*. Collins, London.

Eisenberg, J.F. (1981). *The mammalian radiations. An analysis of trends in evolution, adaptation and behaviour*. The Athlone Press, London.

Elgar, M.A. & Harvey, P.H. (1987). Basal metabolic rates in mammals: allometry, phylogeny and ecology. *Funct. Ecol.* **1:** 25–36.

Frazer, J.F.D. & Huggett, A.St.G. (1974). Species variations in the foetal growth rates of eutherian mammals. *J. Zool., Lond.* **174:** 481–509.

Harvey, P.H. & Mace, G.M. (1982). Comparisons between taxa and adaptive trends: problems of methodology. In *Current problems in sociobiology:* 343–61. (Ed. King's College Sociobiology Group). Cambridge University Press, Cambridge.

Huggett, A.St.G. & Widdas, W.F. (1951). The relationship between mammalian foetal weight and conception age. *J. Physiol., Lond.* **114:** 306–17.

Huxley, J.S. (1932). *Problems of relative growth*. Methuen, London.

Jungers, W.L. (Ed.) (1985). *Size and scaling in primate biology*. Plenum Press, New York & London.

Kihlström, J.E. (1972). Period of gestation and body weight in some placental mammals. *Comp. Biochem. Physiol.* **43A:** 673–9.

Kirkwood, J.K. (1985). The influence of size on the biology of the dog. *J. small Anim. Pract.* **26:** 97–110.

Kleiber, M. (1947). Body size and metabolic rate. *Physiol. Rev.* **27:** 511–41.

Kleiber, M. (1961). *The fire of life: an introduction to animal energetics*. John Wiley, New York.

Leitch, I., Hytten, F.E. & Billewicz, W.Z. (1959). The maternal and neonatal weights of some Mammalia. *Proc. zool. Soc. Lond.* **133:** 11–28.

Leutenegger, W. (1973). Maternal-fetal weight relationships in primates. *Folia primatol.* **20:** 280–93.

Leutenegger, W. (1976). Allometry of neonatal size in eutherian mammals. *Nature, Lond.* **263:** 229–30.

Leutenegger, W. (1977). Neonatal-maternal weight relationship in macaques: an example of intrageneric scaling. *Folia primatol.* **27:** 152–9.

MacDowell, E.C. & Allen, E. (1927). Weight of mouse embryos, 10–18 days after conception, a logarithmic function of embryo age. *Proc. Soc. exp. Biol. Med.* **24:** 672–4.

MacDowell, E.C., Allen, E. & MacDowell, C.G. (1927). The prenatal growth of the mouse. *J. gen. Physiol.* **11:** 57–70.

MacLarnon, A.M., Chivers, D.J. & Martin, R.D. (1986). Gastrointestinal allometry in primates and other mammals including new species. In *Primate ecology and conservation*: 75–85. (Eds Else, J.G. & Lee, P.C.). Cambridge University Press, Cambridge.

MacLarnon, A.M., Martin, R.D., Chivers, D.J. & Hladik, C.M. (1986). Some aspects of gastrointestinal allometry in primates and other mammals. In *Définition et origines de l'homme:* 293–302. (Ed. Sakka, M.). Éditions du CNRS, Paris.

MacMahon, T.A. & Bonner, J.T. (1983). *On size and life.* Scientific American Books, New York.

Martin, R.D. (1975). The bearing of reproductive behavior and ontogeny on strepsirhine phylogeny. In *Phylogeny of the Primates: a multi-disciplinary approach:* 265–97. (Eds Luckett, W.P. & Szalay, F.S.). Plenum Press, New York.

Martin, R.D. (1980). Adaptation and body size in primates. *Z. Morph. Anthrop.* **71:** 115–24.

Martin, R.D. (1981). Relative brain size and basal metabolic rate in terrestrial vertebrates. *Nature, Lond.* **293:** 57–60.

Martin, R.D. (1982). Allometric approaches to the evolution of the primate nervous system. In *Primate brain evolution:* 39–56. (Eds Armstrong, E. & Falk, D.). Plenum Press, New York & London.

Martin, R.D. (1983). *Human brain evolution in an ecological context.* (52nd James Arthur Lecture on the Evolution of the Human Brain.) American Museum of Natural History, New York.

Martin, R.D. (1984a). Scaling effects and adaptive strategies in mammalian lactation. *Symp. zool. Soc. Lond.* No. 51: 87–117.

Martin, R.D. (1984b). Body size, brain size and feeding strategies. In *Food acquisition and processing in primates:* 73–103. (Eds Chivers, D.J., Wood, B.A. & Bilsborough, A.). Plenum Press, New York & London.

Martin, R.D., Chivers, D.J., MacLarnon, A.M. & Hladik, C.M. (1985). Gastrointestinal allometry in primates and other mammals. In *Size and scaling in primate biology:* 61–89. (Ed. Jungers, W.L.). Plenum Press, New York & London.

Martin, R.D. & Harvey, P.H. (1985). Brain size allometry: ontogeny and phylogeny. In *Size and scaling in primate biology:* 147–73. (Ed. Jungers, W.L.). Plenum Press, New York & London.

Martin, R.D. & MacLarnon, A.M. (1985). Gestation period, neonatal size and maternal investment in placental mammals. *Nature, Lond.* **313:** 220–3.

Martin, R.D. & MacLarnon, A.M. (1986). Maternal investment in mammals: reply to Zeveloff and Boyce. *Nature, Lond.* **321:** 538.

McNab, B. (1978). Energetics of arboreal folivores: physiological problems and ecological consequences of feeding on an ubiquitous food supply. In *The ecology of arboreal folivores:* 153–62. (Ed. Montgomery, G.G.). Smithsonian Institution Press, Washington.

McNab, B. (1980). Food habits, energetics, and the population biology of mammals. *Am. Nat.* **116:** 106–24.

McNab, B. (1986). The influence of food habits on the energetics of eutherian mammals. *Ecol. Monogr.* **56:** 1–19.

Melton, D.A. (1976). The biology of aardvarks (Tubulidentata–Orycteropodidae). *Mammal Rev.* **6:** 75–88.

Millar, J.S. (1981). Pre-partum reproductive characteristics of eutherian mammals. *Evolution, Lawrence, Kans.* **35:** 1149–63.

Müller, E.F. (1983). Wärme- und Energiehaushalt bei Halbaffen (Prosimiae). *Bonn. zool. Beitr.* **34:** 29–71.

Müller, E.F. (1985). Basal metabolic rates in primates—the possible role of phylogenetic and ecological factors. *Comp. Biochem. Physiol.* **81A:** 707–11.

Payne, P.R. & Wheeler, E.F. (1967a). Growth of the foetus. *Nature, Lond.* **215:** 849–50.

Payne, P.R. & Wheeler, E.F. (1967b). Comparative nutrition in pregnancy. *Nature, Lond.* **215:** 1134–6.

Payne, P.R. & Wheeler, E.F. (1968). Comparative nutrition in pregnancy and lactation. *Proc. Nutr. Soc.* **27:** 129–38.

Peters, R.H. (1983). *The ecological implications of body size.* Cambridge University Press, Cambridge, London. etc.

Portmann, A. (1938). Die Ontogenese der Säugetiere als Evolutionsproblem. II. Zahl der Jungen, Tragzeit und Ausbildungsgrad der Jungen bei der Geburt. *Biomorphosis* **1:** 109–26.

Portmann, A. (1939). Nesthocker und Nestflüchter als Entwicklungszustände von verschiedener Wertigkeit bei Vögeln und Säugern. *Rev. suisse Zool.* **46:** 385–90.

Portmann, A. (1941). Die Tragzeiten der Primaten und die Dauer der Schwangerschaft beim Menschen: ein Problem der vergleichenden Biologie. *Rev. suisse Zool.* **48:** 511–18.

Portmann, A. (1962). Cerebralisation und Ontogenese. *Medizin. Grundlagenforsch.* **4:** 1–62.

Portmann, A. (1965). Über die Evolution der Tragzeit bei Säugetieren. *Rev. suisse Zool.* **72:** 658–66.

Richard, A.F. & Nicoll, M.E. (1987). Female social dominance and basal metabolism in a Malagasy primate, *Propithecus verreauxi. Am. J. Primatol.* **12:** 309–14.

Robbins, C.T. & Robbins, B.L. (1979). Fetal and neonatal growth patterns and maternal reproductive effort in ungulates and subungulates. *Am. Nat.* **114:** 101–16.

Roberts, C.M. & Perry, J.S. (1974). Hystricomorph embryology. *Symp. zool. Soc. Lond.* No. 34: 333–60.

Ross, C. (1988). The intrinsic rate of natural increase and reproductive effort in primates. *J. Zool., Lond.* **214:** 199–219.

Rudder, B.C.C. (1978). Monkeys as models in medicine. *New Scient.* **77:** 164–6.

Rudder, B.C.C. (1979). *The allometry of primate reproductive parameters*. PhD Thesis: University of London.
Sacher, G.A. & Staffeldt, E.F. (1974). Relation of gestation time to brain weight for placental mammals: implications for the theory of vertebrate growth. *Am. Nat.* **108:** 593–616.
Schmidt-Nielsen, K. (1984). *Scaling: why is animal size so important*? Cambridge University Press, Cambridge, London, etc.
Thompson, A.D. & Nicoll, M.E. (1986). Basal metabolic rate and energetics of reproduction in therian mammals. *Nature, Lond.* **321:** 690–3.
Western, D. (1979). Size, life history and ecology in mammals. *Afr. J. Ecol.* **17:** 185–204.
Western, D. & Ssemakula, J. (1982). Life history patterns in birds and mammals and their evolutionary interpretation. *Oecologia* **54:** 281–90.
Zeveloff, S.I. & Boyce, M.S. (1980). Parental investment and mating systems in mammals. *Evolution, Lawrence, Kans.* **34:** 973–82.
Zeveloff, S.I. & Boyce, M.S. (1986). Maternal investment in mammals. *Nature, Lond.* **321:** 537–8.

Symp. zool. Soc. Lond. (1988) No. 60 81–117

The hormonal and hypothalamic control of primate sexual behaviour

A. F. DIXSON
and S. A. C. LLOYD

MRC Reproductive Biology Unit
Centre for Reproductive Biology
37 Chalmers Street
Edinburgh EH3 9EW, UK

Synopsis

This paper deals with the effects of gonadal and adrenocortical steroids upon the hypothalamus and sexual behaviour in primates. The evidence for hormonal effects upon female sexual initiating behaviour (proceptivity) and sexual attractiveness is discussed. In some species of monkeys and apes these effects are more pronounced than those that concern the female's willingness to allow copulation (sexual receptivity). Receptivity is less dependent upon hormonal control in anthropoid primates than in prosimians or many other mammals. It is suggested that during the evolution of monkeys and apes, sexual selection may have favoured neuroendocrine mechanisms that allow females greater influence or 'choice' in sexual contexts. These mechanisms include hormonal effects upon aspects of female sexual attractiveness (e.g. sexual skin, vaginal cues) and proceptivity (facial displays, body postures, gestures, seeking proximity to the male) as well as a less rigid control of receptivity by ovarian oestrogens and progesterone.

At the hypothalamic level, hormonal effects upon sexual behaviour in monkeys, apes and man differ from those reported for rodents and other mammals. The ventromedial hypothalamus (VMH) is important in the control of sexual receptivity (lordosis) in rodents. However, female anthropoid primates do not exhibit lordotic postures during mating and it remains to be determined whether the VMH influences receptivity in monkeys or apes. In the female common marmoset, anteromedial hypothalamic lesions reduce proceptivity, but receptivity is not usually affected. Exogenous luteinizing hormone releasing hormone (LHRH) stimulates proceptivity in ovariectomized, oestrogen-primed marmosets but has no apparent effect upon receptivity, since, under these conditions, the animals are fully receptive prior to administration of the peptide.

The role of ovarian and adrenal androgens in female sexuality is discussed. In the rhesus monkey, adrenal androgens enhance sexual behaviour by acting upon the anterior hypothalamus. However, in the stumptail macaque and marmoset, adrenalectomy does not affect proceptivity or receptivity in ovariectomized females. The possible role of androgens in women's sexuality is an unresolved question. Since interspecific variability is pronounced, evidence obtained for one primate species (e.g. the rhesus monkey) cannot readily be applied to another (e.g. man).

In male primates, testicular androgens are important for the maintenance of

ZOOLOGICAL SYMPOSIUM No. 60
ISBN 0-19-854007-8

sexual behaviour, as in other mammals. Enlargement of the neocortex during primate evolution has not apparently diminished the role of testicular hormones in regulating sexual behaviour. Lesions in the preoptic area or anterior hypothalamus (POA/AH) cause a striking decrease in copulatory activity in male rhesus monkeys and marmosets, as in other mammals. Electrophysiological studies of macaques suggest that the POA is concerned with precopulatory behaviour, whereas the dorsomedial and posterior hypothalamus form part of the copulatory mechanism (Yoshimatsu 1983). Further studies of primates are required to test this hypothesis.

Visual communication is of great importance for sexual behaviour in monkeys, apes and man, but little is known of the means by which visual stimuli impinge upon hypothalamic mechanisms. Mutual eye contact often forms part of the precopulatory behaviour of both sexes, and facial expressions are used to invite copulation or as distance-reducing displays. Cells in the temporal cortex of monkeys are known to respond specifically to facial features and changes in expression (e.g. Perrett, Smith, Potter, Mistlin, Head, Milner & Jeeves 1985). The temporal cortex has connections to the limbic system (e.g. entorhinal area, amygdala etc.). It may be via such connections that visual stimuli influence the hypothalamic regulation of proceptivity in the female or sexual arousal in the male.

Introduction

Effects of gonadal hormones upon the hypothalamus and sexual behaviour have been documented in amphibians, reptiles and birds (Hutchison 1978; Crews & Silver 1985) and in a variety of mammals (Komisaruk 1978; Larsson 1979; Morali & Beyer 1979; Pfaff 1980; Hart & Leedy 1985). In many species oestrogen, either alone or in combination with progesterone, is required for the full expression of female sexual receptivity or 'oestrus'. In the rat, the ventromedial hypothalamus is an important target for these hormonal effects (Pfaff 1980; Barfield, Rubin, Glaser & Davis 1983). However, the exact sites of steroid action upon receptivity may differ between species (Pfaff & Modianos 1985). Ovarian hormones also influence the female's sexual initiating behaviour (proceptivity: Beach 1976), but the neural basis of proceptivity has received less attention in studies of neuroendocrinology and sexual behaviour. In the anthropoid primates (monkeys, apes and man) ovarian hormones have less rigid effects upon feminine patterns of sexual behaviour than in prosimians and other mammals (Goy & Resko 1972; Herbert 1978; Dixson 1983a). In this review we consider the evidence for central actions of hormones on proceptivity and receptivity in primates and report some recent experiments on the effects of hypothalamic lesions upon proceptivity and receptivity in female monkeys.

The medial preoptic area and anterior hypothalamus (POA/AH) has repeatedly been shown to play a crucial role in masculine sexual behaviour. Destruction of this area causes a prompt decrease in copulatory behaviour in rodents, carnivores and rhesus monkeys (Hart & Leedy 1985). Neurones in the POA bind testosterone, and implantation of testosterone or testosterone propionate into the POA activates sexual behaviour in castrated males (dove: Hutchison 1978; rat: Davidson 1966). The functions of the POA in sexual behaviour are not fully resolved and it seems that they differ between species. For instance, Hansen (1982) suggests that in the rat the POA is concerned primarily with the 'consummatory aspects' of sexual behaviour (copulatory mechanism: Beach 1956) rather than with initiating sexual interactions. In the rhesus monkey, by contrast, there is evidence that the POA is important in sexual arousal whereas the dorsomedial and posterior areas of the hypothalamus form part of the copulatory mechanism (Yoshimatsu 1983). The second purpose of this review is to consider the evidence for a 'dual control' of sexual behaviour by the hypothalamus in male primates and to present new data on behavioural effects of hypothalamic lesions.

Investigations

Hormones and sexual behaviour in female primates

Rhythmic changes in sexual behaviour during ovarian cycles occur in many species of monkeys and apes. In general, ejaculations increase in frequency during the follicular phase or peak during the peri-ovulatory phase of the cycle, and are lower during the luteal phase (Goy & Resko 1972; Herbert 1978; Michael & Bonsall 1979; Dixson 1983a). As an example, changes in sexual behaviour during the post-partum period in six family groups of marmosets are shown in Fig. 1. A video-monitoring system was used to record all sexual interactions for 24–30 days after parturition and blood samples were drawn from females to measure plasma luteinizing hormone (LH) and progesterone (Dixson & Lunn 1987). Frequencies of mounts and ejaculations increased during the post-partum period and were highest during the peri-ovulatory phase (defined as ± 3 days from the pre-ovulatory LH peak). However, no absolute restriction of mating to the peri-ovulatory phase was evident, either in this experiment or in previous studies in which pair-testing techniques were used to examine sexual behaviour during the ovarian cycle (Kendrick & Dixson 1983). The exact pattern of sexual interactions during cycles in primates varies, depending upon the species and on whether sexual behaviour is observed in the wild habitat, in captive social groups, or during paired encounters. Female proceptivity (sexual initiating behaviour) or receptivity (willingness to accept the male) and non-behavioural qualities of female sexual attractiveness may all fluctuate during ovarian cycles. Therefore, an observed change in the male's sexual activity could be due to several factors. In the sections that follow, we consider the evidence for central effects of oestrogens, progestagens and androgens upon proceptivity or receptivity in primates.

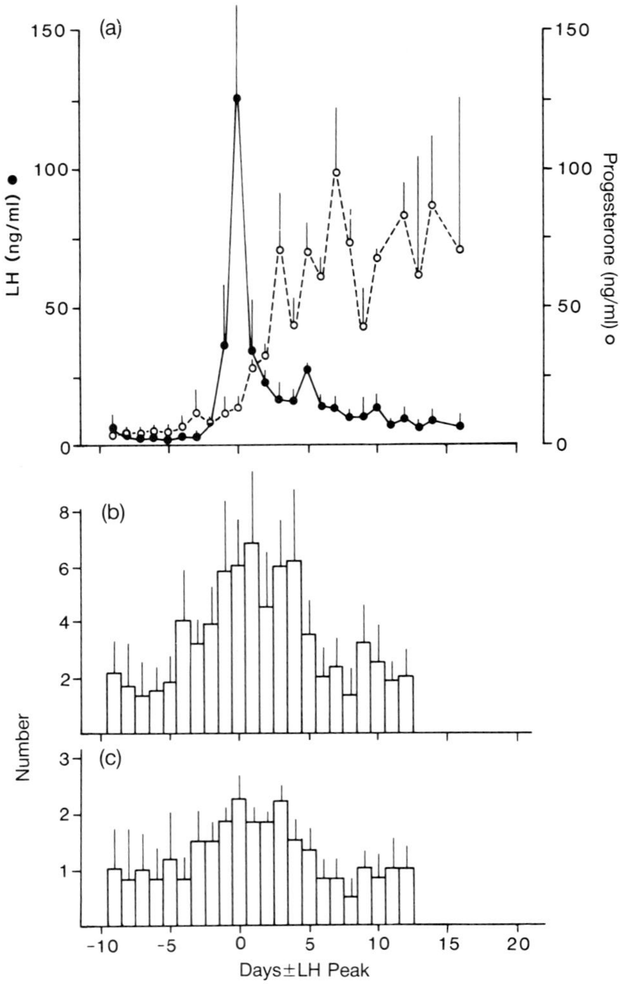

Fig. 1. Changes in hormones and sexual behaviour during the post-partum period in six family groups of marmosets. (a) Overall daily means and standard errors for plasma levels of luteinizing hormone (LH) and progesterone during the post-partum period in six female marmosets. Data are plotted with respect to the occurrence of the LH peak (Day 0). (b) and (c) show overall daily means (+ s.e.m.) for frequencies of males' mounts and ejaculations respectively in these six family groups of marmosets. Data redrawn from Dixson & Lunn (1987).

Oestrogens and sexual behaviour

Oestradiol 17β has pronounced effects upon female sexual attractiveness in some primates, due to its action upon the vagina (rhesus monkey: Herbert 1966; Michael & Keverne 1968) or upon visually attractive 'sexual skin swellings' (Dixson 1983b; chacma baboon: Girolami & Bielert in press). A number of studies have shown that ovariectomy reduces sexual interactions in primates or that oestradiol reverses such effects (e.g. chimpanzee: Young & Orbison 1944; rhesus monkey: Herbert 1970; chacma baboon: Saayman 1973; talapoin: Dixson & Herbert 1977). There is an obvious requirement to control the effects of oestradiol upon female attractiveness if neural effects upon proceptivity or receptivity are to be demonstrated. In practice this may be difficult to achieve. In studies of ovariectomized female rhesus monkeys intra-hypothalamic oestradiol implants caused vaginal cornification, owing to leakage of the hormone into the general circulation (Michael 1969, 1971). The female's sexual behaviour may also be influenced by variables such as the greater size and assertiveness of the male (e.g. orang-utan: Nadler 1982), preferences for particular individuals (pig-tail monkeys: Eaton 1973), or dominant males in social groups (talapoin: Dixson & Herbert 1977). Careful observations of rhesus monkeys during pair tests show that females seek proximity to the male partner and are less 'fearful' (i.e. there is less fear-grimacing) during mid-cycle (Goy & Resko 1972; Czaja & Bielert 1975). Experiments in which males are tethered (Pomerantz & Goy 1983) or the size of the test area is increased (Wallen 1982) indicate that oestrogen may have a more pronounced effect on proceptivity than previously realized. In some species, female invitational behaviour plays a crucial role in determining whether copulation will occur under natural conditions (gorilla: Harcourt, Fossey, Stewart & Watts 1980; orang-utan: Schürmann 1982; brown capuchin: Janson 1984; woolly spider monkey; Milton 1985). It is possible that oestradiol has a greater effect upon proceptivity than upon sexual receptivity in such species. Some of our own experiments on the common marmoset (*Callithrix jacchus*) support this view. In the marmoset the sexes are of equal size and males do not dominate sexual interactions when pair-tested with females in the laboratory. Further, the female is attractive to the male throughout the ovarian cycle or after ovariectomy and adrenalectomy (Kendrick & Dixson 1983, 1984a; Dixson 1987a). Thus, in the restricted conditions of a pair-test, steroid hormones have minimal effects upon genital cues and sexual attractiveness. Administration of oestradiol 17β to ovariectomized marmosets causes a striking increase in proceptive tongue-flicking, crouching, staring and lip-smacking displays (Fig. 2). These increases in proceptivity are presumably due to a central action of oestradiol. Receptivity may also be enhanced slightly by oestradiol (Kendrick & Dixson 1985a) so that females refuse or

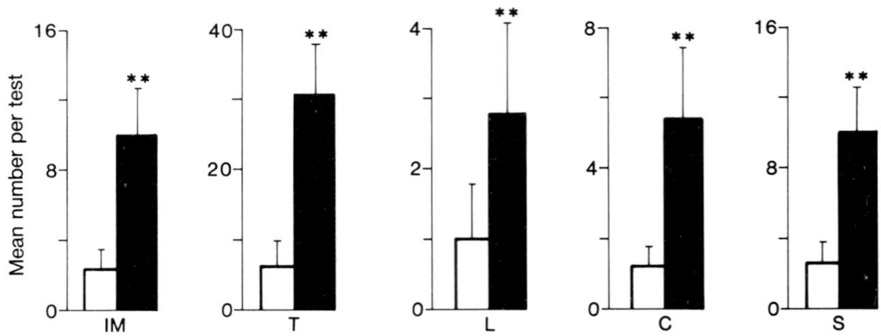

Fig. 2. Mean (+ s.e.m.) frequencies per test of proceptive behaviour—body and limbs immobile (IM), tongue-flicking (T), lipsmacking (L), crouching (C) and stares (S)—shown by nine ovariectomized marmosets before hormone treatment (open bars) and after treatment with oestradiol 17β (closed bars). ★★ $P < 0.01$, Wilcoxon test. Data are from Dixson (1986a).

terminate fewer mounts after hormonal treatment. However, these effects are much smaller than those affecting proceptive behaviour since females may remain receptive after ovariectomy or adrenalectomy (Kendrick & Dixson 1984a; Dixson 1987a).

Progestagens and sexual behaviour

Progesterone reduces female sexual attractiveness in the rhesus monkey by affecting vaginal cues (Baum, Keverne, Everitt & Herbert 1976). It also contributes to the detumescence of the sexual skin swelling during the luteal phase of the menstrual cycle (e.g. chacma baboon: Gillman & Stein 1941) and may by this means influence sexual attractiveness in baboons, mangabeys, mandrills, some macaques, chimpanzees and other species that possess a sexual skin. In a number of mammals progesterone has a facilitatory effect upon sexual receptivity when administered to the 'oestrogen primed' female (e.g. rat, hamster, guinea-pig etc.). This may be followed by an inhibitory effect upon receptivity (Feder 1978). Under certain circumstances progesterone can facilitate a surge of LH in oestrogen-primed rhesus monkeys or women (Pohl & Knobil 1982), but there is no evidence for such a facilitatory action upon sexual behaviour in female primates. However, progestagens may have an inhibitory effect upon sexual behaviour in some women using oral contraceptives (e.g. Grant & Pryse Davies 1968; Bancroft 1983) or when administered in pharmacological dosages to female rhesus monkeys (Michael, Saayman & Zumpe 1968). In ovariectomized common marmosets, subcutaneous implantation of silastic capsules containing progesterone, sufficient to produce plasma levels within the normal range for the luteal phase of the cycle, causes a striking decrease in proceptivity and a lesser decrease in sexual receptivity (Fig. 3). Proceptive 'tongue-flicking'

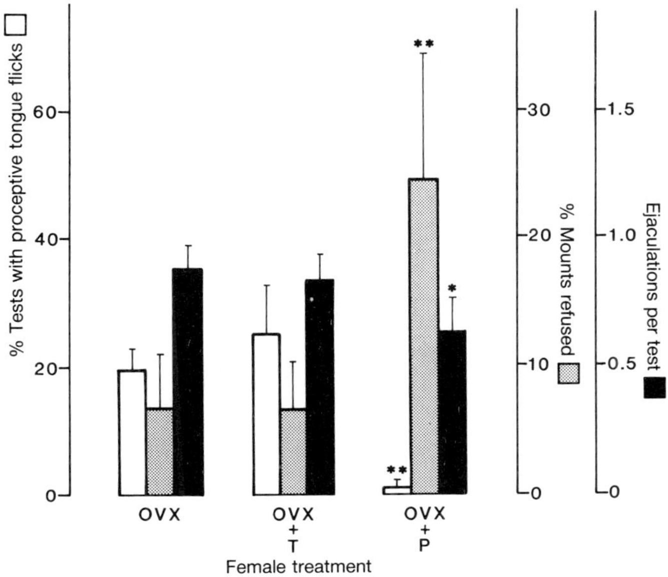

Fig. 3. Mean (+ s.e.m.) percentage of tests with proceptive tongue-flicking (open bars), percentage of mounts refused (dotted bars) and ejaculatory frequencies per test (solid bars) during periods when ovariectomized female marmosets received no hormone treatment (OVX), testosterone (OVX+T) or progesterone (OVX+P). ★ $P < 0.05$, ★★ $P < 0.01$, Wilcoxon test; compared to OVX results. Data for eight pairs of monkeys are from Kendrick & Dixson (1985a).

displays are virtually abolished under these conditions and females refuse, on average, 25% of males' attempts to mount. Intravaginal application of a progesterone-containing cream produces a small increase in circulating hormone concentrations but does not affect sexual behaviour (Kendrick & Dixson 1985a). These experiments indicate that, in marmosets, progesterone has central effects upon proceptivity and to a lesser extent on female sexual receptivity but is unlikely to affect sexual attractiveness during laboratory pair tests. In family groups of marmosets, changes in female attractiveness may be important during the post-partum increase in mating activity (Dixson & Lunn 1987). Both sexes scent-mark by means of urine and secretions from circumgenital cutaneous glands. Future research may demonstrate some relationship between hormones, chemical cues and sexual attractiveness in female marmosets, but these factors appear to be of less importance during laboratory pair-tests.

Androgens and sexual behaviour

Peri-ovulatory increases in circulating testosterone have been measured in a number of primate species (e.g. rhesus monkey: Hess & Resko 1973;

gorilla: Nadler, Graham, Collins & Gould, 1979; owl monkey: Bonney, Dixson & Fleming 1980). This has led to the suggestion that testosterone might influence female sexuality during the peri-ovulatory period (Everitt & Herbert 1972; Dixson, Everitt, Herbert, Rugman & Scruton 1973). Treatment of ovariectomized rhesus monkeys with testosterone propionate leads to an increase in proceptivity (Herbert & Trimble 1967), but large doses of hormone which produce higher than normal plasma levels are required (Michael & Zumpe 1977). Michael, Richter, Cain, Zumpe & Bonsall (1978) treated ovariectomized rhesus monkeys daily with graded doses of oestradiol, progesterone and testosterone in order to produce 'artificial menstrual cycles'. When testosterone was omitted, females continued to show a mid-cycle increase in proceptivity (measured by operant techniques). Thus ovarian androgens are not essential for mid-cycle changes in proceptivity in rhesus monkeys.

Since the adrenal glands also secrete androgens and other 'sex steroids' it is possible that these might influence female receptivity or proceptivity. In a series of experiments Everitt, Herbert and their colleagues showed that adrenal androgens enhance proceptive and receptive behaviour in rhesus monkeys (Everitt & Herbert 1971; Everitt, Herbert & Hamer 1972). Thus adrenalectomy or adrenocortical suppression by means of dexamethasone caused a decrease in sexual behaviour which was reversed by treatment with testosterone or androstenedione. With different testing methods Johnson & Phoenix (1976) found that decreases in proceptivity occurred when female rhesus monkeys received dexamethasone, but the changes were not statistically significant. Wallen & Goy (1977) reported that dihydrotestosterone propionate (DHTP) has no effect on proceptivity in ovariectomized rhesus monkeys whereas testosterone propionate (TP) and oestradiol benzoate enhance proceptivity. Since DHTP is not 'aromatized' to oestrogen within the brain it is possible that only androgens which are converted to oestrogen centrally can stimulate proceptivity in rhesus monkeys. However, in the experiments by Everitt *et al.* cited above, ovariectomized females received oestradiol benzoate continuously in order to maintain their sexual attractiveness to males. It seems unlikely that subsequent androgen treatment stimulated proceptivity simply by augmenting central concentrations of oestrogen.

Unilateral implants of TP into the anterior hypothalamus of ovariectomized, adrenalectomized rhesus monkeys enhance proceptivity and also increase receptivity, as reflected by a decrease in refusals of males' mounting attempts in some females (Everitt & Herbert 1975). Binding of tritiated testosterone in this region of the brain has been demonstrated by autoradiography (Sheridan, Hagino & Weaker 1982). Unfortunately the work by Everitt & Herbert (1975) is the only controlled study of intrahypothalamic steroidal effects upon sexual behaviour in a primate species. Nor should the findings in rhesus monkeys be applied uncritically to other primates. In women some evidence for effects of androgens upon sexuality has been obtained but much doubt remains (Bancroft 1983). In stumptail macaques dexamethasone treatment or adrenalectomy has no effect on sexual behaviour in ovariectomized females (Baum, Slob, de Jong & Westbroek 1978). In the common marmoset, although oestradiol 17β has a major stimulatory effect on proceptivity, neither exogenous testosterone nor adrenalectomy has any measurable effects on sexual interactions during laboratory pair-tests (Figs 3 and 4).

Effects of hypothalamic lesions upon proceptivity and receptivity

In the rhesus monkey, extensive hypothalamic lesions do not abolish the menstrual cycle, provided that the medial basal region and arcuate nucleus remain intact (Pohl & Knobil 1982). The control of luteinizing hormone secretion in rhesus monkeys therefore differs from that observed in some non-primates (e.g. rat: Kalra & Kalra 1983; sheep: Martin 1984). It is likely that the hypothalamic control of sexual behaviour also differs in anthropoid primates, given that effects of ovarian hormones upon behaviour are less rigid than in many other mammals. In the ovariectomized rat, lesions of the ventromedial hypothalamus cause striking reductions in the proceptivity and receptivity normally elicited by treatment with oestradiol and progesterone (Edwards & Matthews 1977; Pfaff 1980). Lesions of the anterior hypothalamus cause decreases in receptivity in the rat (Law & Meagher 1958), hamster (Malsbury, Kow & Pfaff 1977) and sheep (Clegg, Santolucito, Smith & Ganong 1958). However, lesions in some areas may enhance the display of lordosis in rats (medial preoptic area: Powers & Valenstein 1972; septum: Nance, Shryne & Gorski 1975).

The results of the first experiments to examine the effects of anteromedial hypothalamic lesions upon sexual behaviour in female primates are shown in Fig. 5. In this study bilateral lesions were produced in eight ovariectomized, oestradiol-treated marmosets by means of a temperature-controlled electrode. Three females received sham lesions as a control procedure (Kendrick & Dixson 1986). The lesions affected the anterior hypothalamus (Fig. 6A) but also extended into the medial hypothalamus to damage the ventromedial, dorsomedial and paraventricular nuclei to varying extents. All females showed a striking reduction or abolition of proceptivity as a result of these lesions. Five females continued to show proceptive staring, but at greatly reduced frequencies, and only one animal continued to show tongue-flicking behaviour. Seven of the eight females continued to accept normally the mounting attempts of their male partners and receptivity was unaffected. The frequencies of tongue-flicking during intromission were, however, much reduced (Kendrick & Dixson 1986). In this experiment one

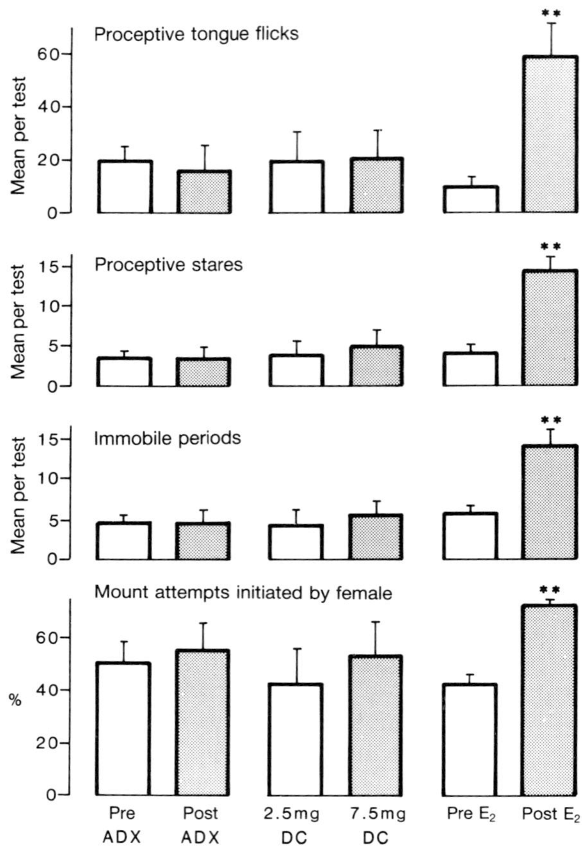

Fig. 4. Effects of adrenalectomy (ADX), deoxycorticosterone (DC) and oestradiol 17β (E_2) upon proceptivity in ovariectomized marmosets. Data are means (+ s.e.m.) per test for six females (20 tests before and 20 tests after ADX; 10 tests before and 10 after E_2) and seven females (10 tests before and 10 tests after the DC dosage increase from 2.5 to 7.5 mg/week). ADX females were maintained on twice-daily injections of 6 mg of hydrocortisone hemisuccinate. ★★ $P = 0.025$, Wilcoxon test. Pre-condition (open bars) and post-condition (dotted bars) for each comparison. Data are from Dixson (1987a).

female (No. 7), with lesions that primarily affected the medial hypothalamus, showed a significant decrease in sexual receptivity.

Effects of medial hypothalamic lesions upon proceptivity and receptivity were examined in a separate experiment. Seven ovariectomized–adrenalectomized females were used, each having received subcutaneous implants of two silastic capsules containing oestradiol 17β. High frequencies of proceptive behaviour are maintained by oestradiol under these conditions (Dixson 1987a), just as in adrenal-intact females. Thermal lesions were produced

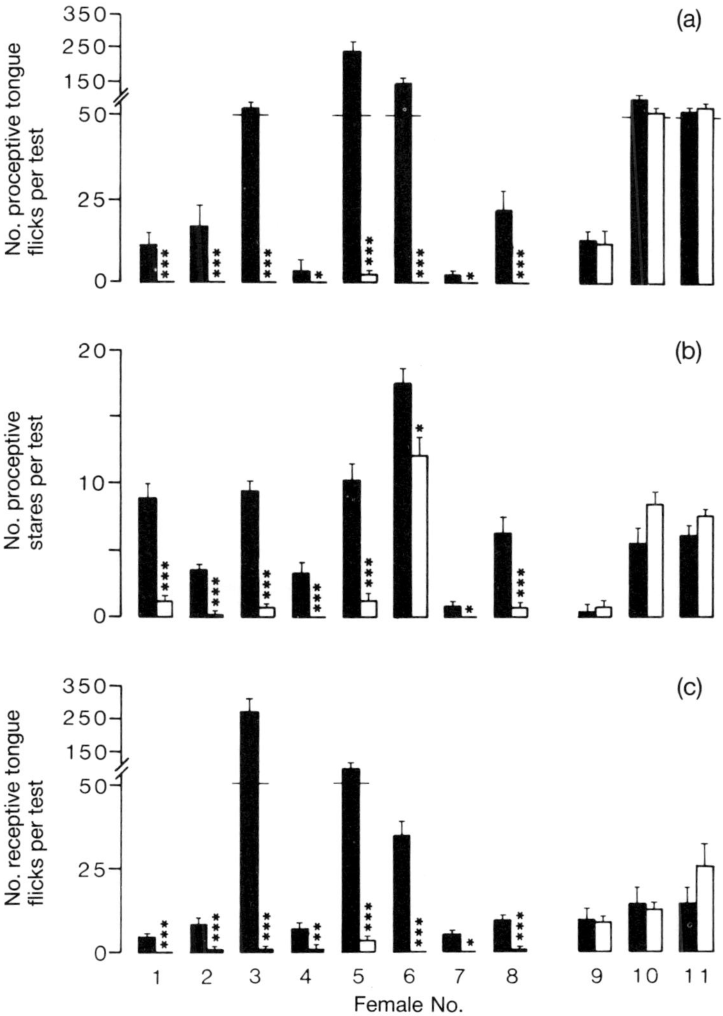

Fig. 5. Effects in marmosets of thermal lesions (females Nos. 1–8) or sham lesions (females Nos. 9–11) of the anteromedial hypothalamus upon frequencies (mean + s.e.m.) of (a) proceptive tongue-flicking, (b) proceptive stares, and (c) receptive tongue-flicking. Data are for each animal before (mean of 10 mating tests) and after lesioning (mean of 20 tests). ★ $P < 0.05$, ★★ $P < 0.01$, ★★★ $P < 0.001$; Mann-Whitney U-test. Solid bars: before lesioning. Open bars: after lesioning. Data are from Kendrick & Dixson (1986).

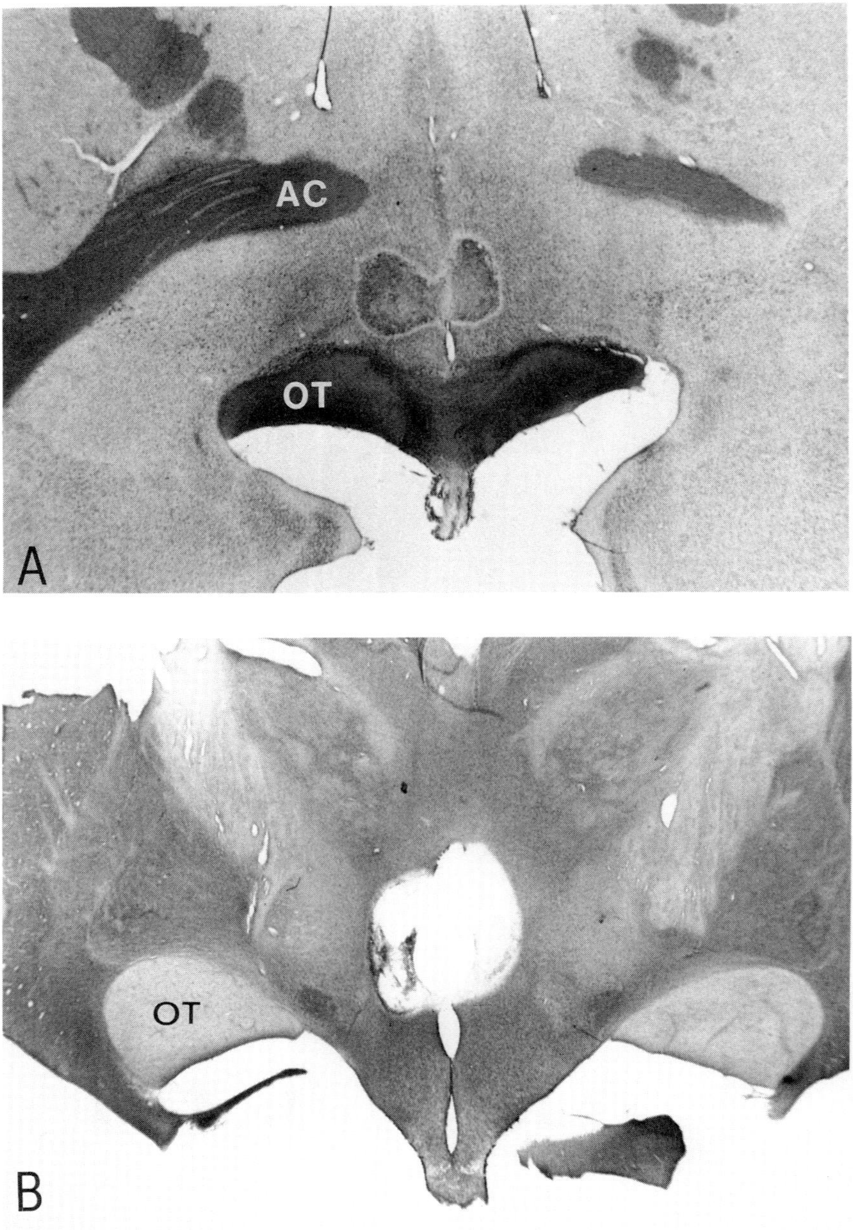

Fig. 6. Transverse sections of the brains of female common marmosets to show lesions in (A) the anterior hypothalamus and (B) the medial hypothalamus. Effects of such lesions upon proceptivity and receptivity are described in the text. AC = anterior commissure, OT = optic tract. Stain = cresyl violet.

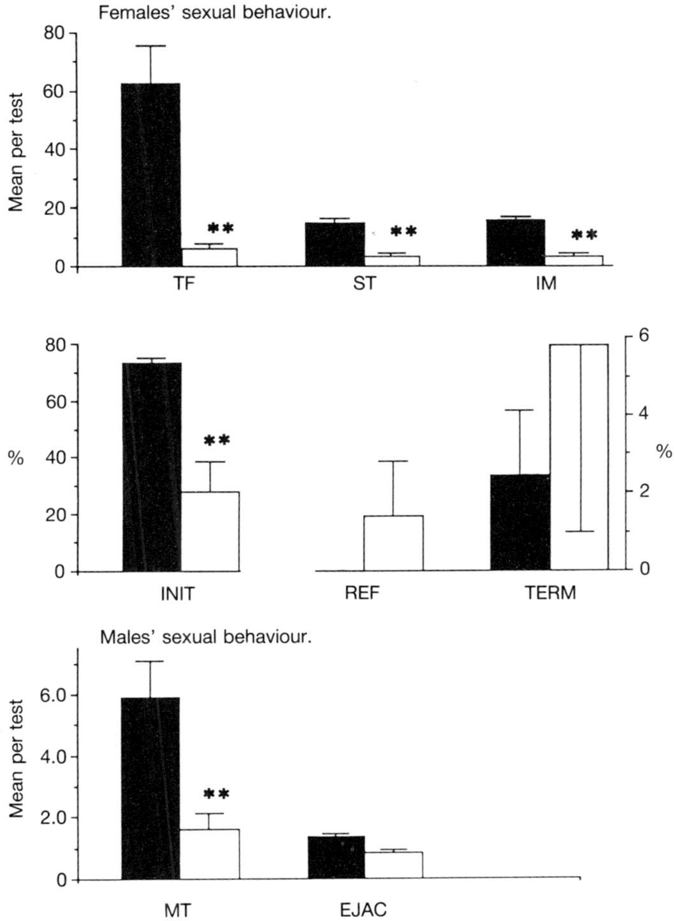

Fig. 7. Effects of medial hypothalamic lesions in ovariectomized, adrenalectomized, oestradiol-treated marmosets upon patterns of copulatory behaviour. Data are means (+ s.e.m.) per test for seven pairs of animals for 10 tests before lesioning (closed bars) and 10 tests after lesioning (open bars). TF = proceptive tongue-flicks, ST = stares, IM = immobile periods, INIT = initiation of males' mounts or mount attempts, REF = refusals of males' mounting attempts, TERM = termination of males' mounts, MT = mount, EJAC = ejaculation. ★★ $P = 0.02$, Wilcoxon test. Data are from A. F. Dixson (unpublished).

bilaterally at two depths, by means of an electrode maintained at 55°C for 15 s at each position. These procedures mainly affected the area dorsalis hypothalami, and the paraventricular and dorsomedial nuclei (Fig. 6B). Partial damage to the dorsal part of the ventromedial nuclei occurred in four animals. All females showed a decrease in proceptive behaviour after the medial hypothalamic lesions had been produced, and they initiated fewer attempts at mounting by their male partners (Fig. 7). Indeed, a significant decrease in the frequency of mounting occurred during post-operative tests; ejaculation also declined in frequency, but this finding was not statistically significant. Figure 7 also shows that these changes in sexual activity were not associated with a reduction in sexual receptivity, as reflected by measurements of refusals or mount terminations by females. During the experiment six females never refused the males' mounting attempts and four never terminated a mount. The remaining females showed no significant changes in receptivity as a result of hypothalamic lesions. Decreases in proceptivity were striking, but only one female showed a complete absence of proceptivity after medial hypothalamic lesions had been produced. Some caution must be exercised because no control (sham-lesioned) females have been studied and some decline in proceptivity might have occurred owing to the effects of repeated testing under continuous exposure to oestradiol.

These experiments indicate that a distinction may be made between the hypothalamic control of proceptivity and receptivity in primates. Anteromedial lesions have a much greater effect upon proceptive behaviour than on receptivity. It remains to be determined whether for any primate species the ventromedial hypothalamus is concerned in the control of sexual receptivity. So far, only partial destruction of this area has been achieved by means of thermal lesions in the marmoset.

Neurotransmitters and female sexuality: effects of LHRH

The ability of luteinizing hormone releasing hormone (LHRH) to potentiate sexual receptivity was first demonstrated in the rat. The lordotic reflex is enhanced in ovariectomized, oestrogen-primed rats by administering LHRH peripherally (Moss & McCann 1973) or by infusing the decapeptide into hypothalamic sites (Moss & Foreman 1976), the mid-brain central grey (Sakuma & Pfaff 1980), or the vertebral canal (Sirinathsinghji 1983). Although extensive networks of LHRH-containing neurones have been mapped in the brains of primates (e.g. man: Barry 1976; squirrel monkey: Barry 1978; baboon: Marshall & Goldsmith 1980) we have little knowledge of the functions of LHRH aside from its classical role as a releasing factor for LH. Some clinical evidence points to an effect of LHRH upon sexual arousal in men (e.g. Benkert, Jordan, Dahlen, Schneider & Gammel 1975; Evans & Distiller 1979). Intramuscular injection of LHRH (25 μg) has also been shown to increase proceptivity in female marmosets (Kendrick & Dixson 1985b). Small silastic capsules (6–16 mm long) containing oestradiol were implanted subcutaneously into eight ovariectomized females in order partly to stimulate proceptive tongue-flicking and staring displays. When LHRH was administered to these females, a further activation of proceptivity was seen within 2 h (Fig. 8). Control injections of saline had no effect upon behaviour, and plasma concentrations of cortisol and progesterone

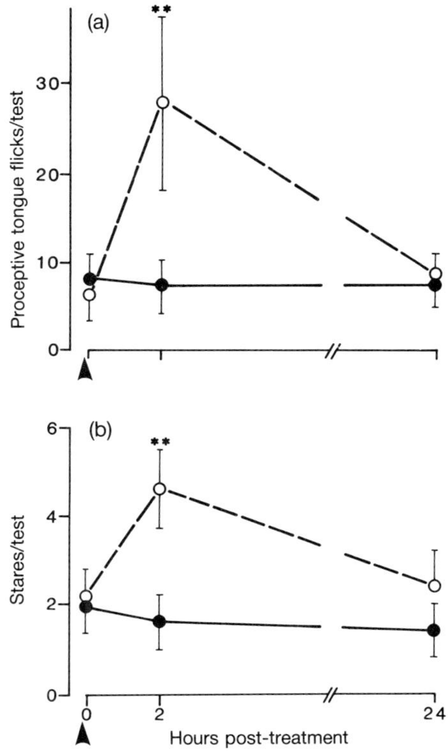

Fig. 8. Effects of LHRH (25 μg intramuscularly, dashed line) or saline (solid line) upon (a) proceptive tongue-flicking, and (b) proceptive stares, by ovariectomized, oestrogen-primed marmosets. $P < 0.01$ (Wilcoxon test) for comparisons between tests at time 0 (before LHRH treatment, indicated by an arrow) and 2 h after LHRH injection. Data are from Kendrick & Dixson (1985b).

did not change significantly after LHRH treatment. Therefore, it is likely that LHRH stimulated proceptivity in these female monkeys by acting upon the brain rather than via changes in adrenocortical steroid secretion. This is the first demonstration that LHRH may influence proceptivity in a primate species, but it remains to be seen whether the peptide normally has such effects, for instance during the peri-ovulatory phase of the ovarian cycle. In the rat, the number of neurones in the ventromedial nucleus of the hypothalamus that respond to mechanical stimulation of the vagina and cervix increases during oestrogen/progesterone treatment. These neurones are also responsive to iontophoretic application of LHRH (Chan, Dudley & Moss 1984). Rising levels of oestradiol prior to ovulation in the marmoset might therefore increase the number (or sensitivity) of LHRH responsive neurones that influence proceptive behaviour.

Hormones and sexual behaviour in male primates

The effects of castration or androgen replacement upon sexual behaviour in primates (rhesus monkey: Michael & Wilson 1974; Phoenix 1973; talapoin: Dixson & Herbert 1977; man: Bancroft 1983) are similar to those reported for other mammals (Larsson 1979). Ejaculation and intromission are the first patterns that decline after castration, whereas various forms of 'sexual interest', such as sniffing or looking at the female genitalia, persist for much longer. Retention of copulatory behaviour after castration in rhesus monkeys (Phoenix 1973) or stumptail macaques (Schenck & Slob 1986) is not exceptional in the context of other mammals such as the beagle (Beach 1970) or Sokoto goat (Hart & Jones 1975). Enlargement of the neocortex during primate evolution has not necessarily been associated with a diminished role for testicular hormones in the control of copulatory behaviour.

Autoradiographic studies have shown that particular regions of the hypothalamus, septum, amygdala, stria terminalis and hippocampus differ from other areas of the brain by binding tritiated testosterone in higher concentrations (e.g. rat: Sar & Stumpf 1973; rhesus monkey: Rees & Michael 1982). Oestradiol is the major nuclear metabolite of testosterone in the amygdala, preoptic area and hypothalamus of the male rhesus monkey, and oestradiol binding sites have been mapped in the brains of castrated males by Bonsall, Rees & Michael (1986). However, there is no evidence that aromatization of testosterone to oestradiol by the brain has significant effects upon sexual behaviour in this species (Phoenix 1974; Phoenix & Chambers 1982).

Implantation of crystalline testosterone propionate into the medial preoptic area and anterior hypothalamus (POA/AH) partly restores sexual behaviour in castrated rats (Davidson 1966). Implants in some other hypothalamic sites are also effective (Smith, Damassa & Davidson 1977). Unfortunately, no comparable studies have been carried out in primates. The involvement of the hypothalamus in sexual arousal and copulatory behaviour of monkeys has been studied by electrophysiological and lesioning techniques. These studies indicate that other hypothalamic areas besides the POA/AH may be involved in the control of masculine copulatory behaviour.

Electrophysiological studies of the hypothalamus and sexual behaviour

MacLean & Ploog (1962) mapped the cerebral representation of penile erection in isolated, restrained squirrel monkeys. Electrical stimulation of the POA, paraventricular nucleus, supraoptic nucleus and lateral hypothalamus all caused penile erection as did stimulation of a number of sites outside the hypothalamus. It is important to note that squirrel monkeys employ penile erections as part of a ritualized 'genital display' in a variety of social contexts and not solely during sexual activity. Erections occur in male marmosets during stimulation of the ventromedial region of the hypothalamus (Lipp & Hunsperger 1978) and stimulation of the POA causes erection and seminal discharge in isolated male rhesus monkeys (Robinson & Mishkin 1966) and rats (Van Dis & Larsson 1970). In the latter studies, electrical stimulation of the caudate, globus pallidus, occipital cortex and various other sites also caused seminal emission in rats so that the physiological significance of the observations is doubtful. More relevant to an understanding of complete patterns of sexual activity are studies of electrical stimulation in unrestrained males. Exaggerated sexual behaviour occurs if a small region encompassing the POA (0.43 mm^3) is stimulated in male rats (Merari & Ginton 1975). The animals exhibit repeated ejaculations as well as shortening of the latent period before ejaculation and of the inter-ejaculatory interval. 'Stimulus bound' effects upon sexual behaviour in male rhesus monkeys have also been demonstrated (Perachio, Marr & Alexander 1979). However, the effective stimulation sites differed from those in the rat. When male rhesus monkeys were tested with oestradiol-treated females, stimulation of the POA caused males to mount for longer periods and to make more pelvic thrusts than during normal copulation. Stimulation of the POA failed to produce ejaculation, however, despite previous findings in restrained males (Robinson & Mishkin 1966). Perachio *et al.* (1979) found that complete patterns of copulatory behaviour in which there were increases in the duration of mount, thrusts per mount, thrust rate, and ejaculatory frequency, were elicited by stimulation of the dorsomedial hypothalamus. In isolated males, stimulation of the anterior, ventromedial and posterior hypothalamus produced penile erection, whereas these sites had no effect on sexual behaviour when males were paired with females.

Yoshimatsu (1983) and Oomura, Yoshimatsu & Auo (1984) carried out elegant studies on the effects of sexual activity upon single unit recordings from hypothalamic neurones in male macaques. An operant paradigm was used to measure male sexual arousal. The male was restrained within sight of the female and could obtain access to her by pressing a bar. Under these conditions POA neurones showed a high firing rate which decreased after acquisition of the female and during copulation. Some POA neurones showed a transient increase in activity in relation to penile erection before copulation occurred. By contrast, neuronal activity in the dorsomedial and posterior hypothalamus increased during mounting, intromission and pelvic thrusting. On this basis Yoshimatsu (1983) proposed that a 'dual control mechanism' exists in the hypothalamus of male monkeys. The POA may be concerned with sexual arousal and precopulatory behaviour primarily, and the dorsomedial and posterior hypothalamus with the 'copulatory mechanism'.

Evidence from lesion studies of the hypothalamus

Electrolytic lesions of the POA suppress sexual activity in males of numerous species, including frogs (Schmidt 1968), rats (Heimer & Larsson 1967), cats (Hart, Haugen & Peterson 1973), dogs (Hart 1974), goats (Hart 1986) and rhesus monkeys (Slimp, Hart & Goy 1978). Such effects are not due to impairment of pituitary-testicular control. The testes of POA-lesioned males remain normal and circulating testosterone is adequate to maintain sexual behaviour. Moreover, POA lesions cause immediate deficiencies in copulatory behaviour, in contrast to the more gradual decline that follows castration. Nor can the effects of POA lesions be explained solely on the basis that males show no sexual interest in females. Hart *et al.* (1973) observed that most male cats with POA lesions sniffed the female's nose and genitalia but then appeared to 'ignore her'. Five out of eight lesioned males continued to mount females at greatly reduced frequencies, without the 'treading' and pelvic thrusting patterns that normally precede intromission. Comparable experiments on beagles (Hart 1974) showed that some males continued to investigate the female's anogenital region, mount and make shallow thrusting movements but then appeared to lose interest. Male mice with POA lesions continued to emit precopulatory ultrasonic vocalizations (Bean, Nunez & Conner 1981) and male goats still exhibited the flehmen response to female urine (Hart 1986). Male rats bearing POA lesions continued to follow females and investigate the anogenital region (Heimer & Larsson 1967). Hansen, Kohler, Goldstein & Steinbusch (1982) selectively destroyed neuronal cell bodies in the POA of male rats with ibotenic acid and reported similar effects; but ibotenic acid lesions in the lateral hypothalamus did not affect copulatory behaviour. Such lesions spare the fibres of the medial forebrain bundle (MFB) which play a crucial role in masculine copulatory behaviour. Thus knife cuts of the MFB which sever connections with the POA strikingly impair the initiation of copulatory behaviour. By contrast, knife cuts dorsal to the POA in rats mainly influence the ability of males to achieve ejaculation (Szechtman, Caggiula & Wulkan 1978).

No studies have been undertaken in primates to examine the effects of neurotoxic lesions or selective knife cuts of POA connections upon masculine sexual behaviour. However, radiofrequency lesions in the POA/AH of male rhesus monkeys caused decreases in sexual activity as in other mammals (Slimp *et al.* 1978). Monkeys with such lesions continued to have erections and to masturbate to ejaculation in their home cages, indicating that the POA/AH is not required to sustain such behaviour. Male baboons masturbate in response to the sight of an attractive female (Bielert, Howard-Tripp & Van Der Walt 1980). It is uncertain whether the lesioned male rhesus monkeys studied by Slimp *et al.* (1978) were exhibiting 'sexual motivation' or 'arousal' during masturbation, since no females were present and the animals habitually behaved in this way when in isolation. However, such observations help to answer one caution regarding POA lesions raised by Pfaff (1980) and Pfaff & Modianos (1985). Pfaff points out that the POA is important in regulating various autonomic functions, and that decreases in sexual activity after lesioning might be due to impairment of the parasympathetic control of erection. Pulmonary oedema due to systemic vasoconstriction can also occur after lesions of the POA, and such effects might influence the males' copulatory behaviour. However, erection is not necessarily impaired by lesions of the POA and electrical stimulation of this area activates sexual responses in species that lack a penis and in which external fertilization is the norm (green sunfish: Demski, Bauer & Gerald 1975; leopard frog: Wada & Gorbman 1977a). In the context of vertebrate evolution the involvement of the POA in the integration of male sexual behaviour pre-dated the evolution of the mammalian penis.

Recently, we have examined the effects of hypothalamic lesions upon sexual behaviour in the male common marmoset (S. A. C. Lloyd & A. F. Dixson, unpublished observations). Six males were paired with females for behavioural testing 20 times before and after the production of bilateral thermal lesions in the hypothalamus. Three sham-operated males served as control subjects. All lesioned males showed profound decreases in sexual activity whereas no significant effects were measured in the controls (Fig. 9). Some individual variation in the behavioural effects of these lesions was apparent. Lesions of the POA and anterior hypothalamus which damaged tissues below the anterior commissure (Fig. 10B) had the most profound effects upon mounting behaviour (e.g. male No. 6, Fig. 11). Similar findings were reported by Slimp *et al.* (1978) for the rhesus monkey. However, male No. 1, in which the lesion had damaged the ventromedial hypothalamus (Fig. 10A), continued to show a high frequency of mounting, with pelvic thrusting. This male rarely attained intromission although his mounts were well orientated and erections appeared normal (Fig. 11). When intromission did occur the male ejaculated, indicating that this part of the copulatory mechanism was still operative.

Male marmosets bearing anteromedial hypothalamic lesions showed increased frequencies of allogrooming and grooming invitational behaviour. Similar findings were reported by Slimp *et al.* (1978) for the rhesus monkey and it appears that such lesions are not associated with a general decline in social behaviour. Male marmosets exhibit a variety of precopulatory patterns which may be indicative of sexual arousal. They may exhibit penile erection prior to introduction of the female into the test cage. Males also 'tongue-flick' prior to mating in a manner similar to that of proceptive females, and they may follow females and investigate their genitalia. Decreases in these behavioural patterns occurred as a result of anteromedial

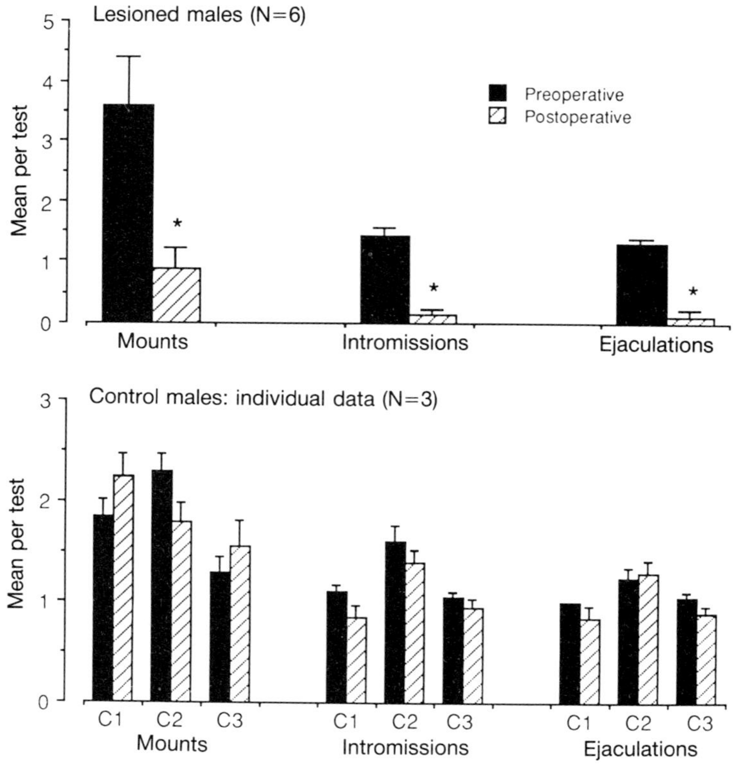

Fig. 9. Effects of antemedial hypothalamic lesions upon sexual behaviour in male marmosets. Data are means and standard errors for 20 tests on six males before lesioning (closed bars) and after lesioning (hatched bars). Individual data on three control (sham-lesion) males are also shown. ★ $P = 0.05$, Wilcoxon test. (S. A. C. Lloyd & A. F. Dixson, unpublished data).

hypothalamic lesions, but some degree of sexual interest persisted (Table 1). Control males also showed decreases in precopulatory patterns, which may indicate that repeated testing with a familiar partner influences the male's sexual interest. Care was taken to keep partners in visual isolation between tests since prolonged visual contact can decrease potency in male rhesus monkeys (Zumpe & Michael 1984).

Lesions of the POA/AH apparently damage linkages between the arousal and copulatory mechanisms in rhesus monkeys. POA/AH lesions have similar effects in the male marmoset. It is possible that additional control of copulatory patterns is exerted by neurones in the dorsomedial hypothalamus or posterior hypothalamus as suggested by Yoshimatsu (1983). Further studies of the effects of lesions in these sites would be valuable. In man unilateral lesions extending from the ventromedial region to

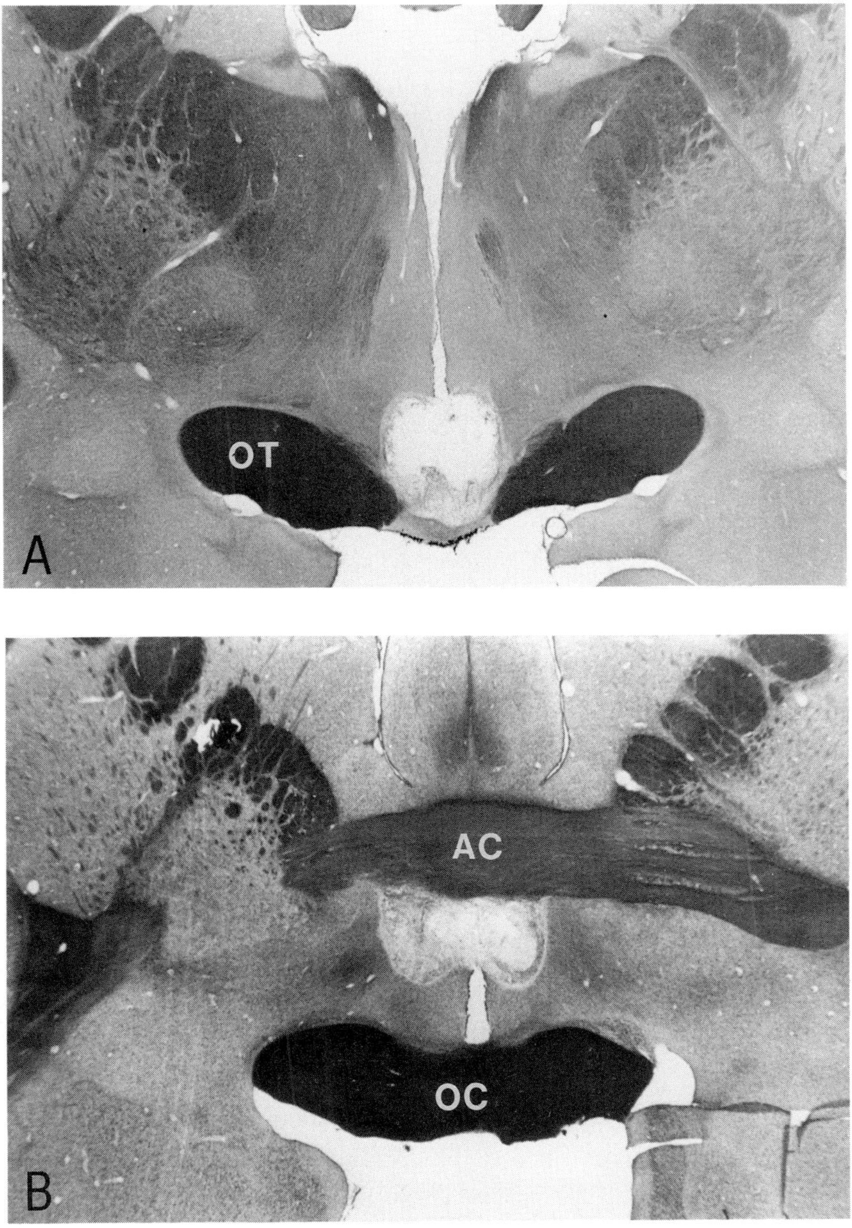

Fig. 10. Transverse sections of the brains of male common marmosets to show lesions in (A) the ventromedial hypothalamus (male No. 1) and (B) the anterior hypothalamus (male No. 6). Effects of these lesions upon sexual behaviour are described in the text and shown in Fig. 11. AC = anterior commissure, OC = optic chiasm, OT = optic tract. Stain = cresyl violet and luxol fast blue.

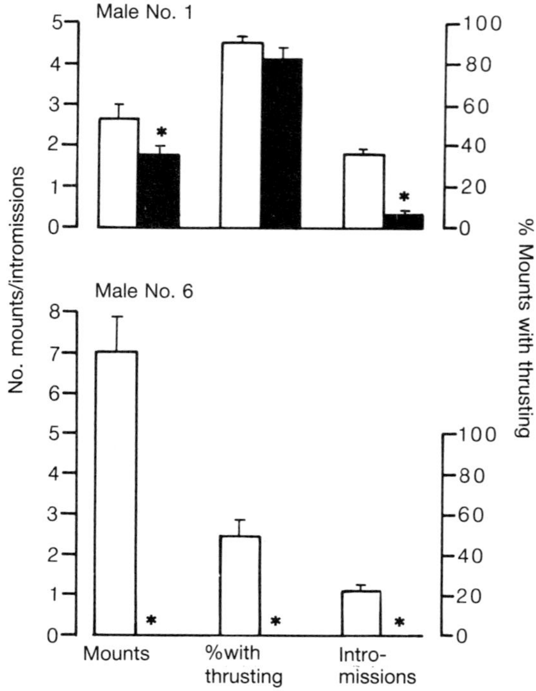

Fig. 11. Effects of thermal lesions in the ventromedial hypothalamus (male No. 1) and anterior hypothalamus (male No. 6) upon mounting, pelvic thrusting and intromission in male marmosets. Data are means and standard errors for 20 tests before (open bars) and after (closed bars) lesioning. Fig. 10A and B shows the positions of lesions in the brains of these two animals and behavioural effects are discussed in the text. $P < 0.05$, Mann Whitney U-test. (S. A. C. Lloyd & A. F. Dixson, unpublished data).

the POA in paedophilic homosexuals and 'hypersexual' men resulted in reductions in sexual interest; and bilateral lesions which may have damaged the anterior hypothalamus and MFB in heterosexual men (Meyers 1961, 1962) led to complete loss of sexual interest (see Hart & Leedy 1985 for a review).

Influences that impinge upon hypothalamic mechanisms

Larsson (1979) has described the POA as 'the origin of a final common pathway for sexual behaviour that is activated by a variety of arousing stimuli from the olfactory and vomeronasal systems; the gustatory, auditive, visual and tactile sensory systems and circulatory hormones.' The importance of various 'arousing stimuli' and the part played by hormones in modulating inputs to the hypothalamus must vary in different vertebrates. Rapid effects of testosterone upon responsiveness of POA neurones to

Table 1. Effects of anteromedial hypothalamic lesions upon sexual arousal and copulatory behaviour in male marmosets.

Animals	Post-operative totals expressed as % of pre-operative totals in respect of				
	mounts	intromission	anticipatory erection	precopulatory tongue flicks	anogenital investigation
Lesioned males					
1	64.8[a]	16.7[a]	85.0	94.4	65.3[a]
2	59.3[a]	39.3[a]	22.2	147.7	105.6
3	33.3[a]	0[a]	5.0[a]	10.1[a]	92.3
4	9.7[a]	0[a]	0[a]	3.4[a]	31.9[a]
5	4.3[a]	0[a]	16.7[a]	0[a]	51.9[a]
6	0[a]	0[a]	0[a]	0[a]	41.3[a]
Control males					
1	121.6	77.3	—[b]	45.0	69.0
2	78.3	87.5	105.9	67.7	61.7
3	119.2	90.5	100.0	83.3	25.5

Data are from 20 tests before and 20 tests after lesioning or sham-lesioning.
[a] $p < 0.05$ Mann Whitney U-test.
[b] This male did not exhibit the behaviour either before or after sham-lesioning (S. A. C. Lloyd & A. F. Dixson, unpublished data).

olfactory stimuli have been described by Pfaff & Pfaffman (1969) in electrophysiological studies of castrated male rats. Such effects are more likely to be important in rodents or prosimians than among anthropoid primates. Tactile stimuli from the penis are important for normal co-ordination of copulatory behaviour in a variety of mammals including primates (rhesus monkey: Herbert 1973; marmoset: Dixson 1986b). Changes in electroencephalographic activity in the POA and in some other hypothalamic areas occur in response to tactile stimulation of the penis in rhesus monkeys; and testosterone propionate (TP) alters such activity within one hour of administration (Mangat, China, Singh & Anand 1978). Hence, testicular hormones may have central effects upon the processing of such tactile stimuli, as well as upon penile morphology (Herbert 1974).

When testosterone is administered to castrated males, some days elapse before sexual behaviour is stimulated. Implantation of testosterone or TP into the POA of castrates restores sexual activity gradually (leopard frog; 3–6 days: Wada & Gorbman 1977b; capon, 4–7 days: Barfield 1969; rat, 5–6 days: Davidson 1966). Rapid effects of testosterone upon responsiveness of POA neurones to olfactory or tactile inputs do not, by themselves, account for changes in sexual behaviour. Hutchison (1978) suggested that different populations of neurones that play a part in sexual behaviour may vary in their sensitivity to testosterone. Experiments by Kendrick and his colleagues with castrated male rats support this hypothesis. Neurones in the stria terminalis, projecting from the corticomedial amygdala to the POA, show increased refractory periods after castration and are restored to normal by TP treatment for seven days. This time course correlates with restoration of the ejaculatory response during copulation (Kendrick, Drewett & Wilson 1981). Lesions of the corticomedial amygdala or stria terminalis disrupt ejaculatory responses in intact male rats (Harris & Sachs 1975). The same is true if knife cuts are made dorsal to the POA, severing connections with the amygdala and hippocampus (Szechtman *et al.* 1978). Kendrick (1983) has also shown that neurones projecting *from* the POA into the medial forebrain bundle (MFB) are testosterone-sensitive. However, only five days are required to restore neuronal refractory periods in castrated males, which correlates with restoration of mounting and intromission. These components of sexual behaviour are abolished or greatly reduced by knife cuts of the MFB in male rats (Paxinos & Bindra 1971).

Visual communication is of vital importance in the social lives of monkeys, apes and man, so we should expect visual stimuli to play a crucial role in sexual contexts. However, there is little knowledge of how such stimuli impinge upon hypothalamic mechanisms. Kluver & Bucy (1939) first demonstrated that removal of the temporal lobes (including the amygdala) in male rhesus monkeys resulted in hyper-oral and inappropriate sexual behaviour. Mounting of inanimate objects also occurs after temporal lobe damage in the cat (Green, Clemente & De Groot 1957). It is interesting that learning of visual tasks in macaques depends on processing of visual inputs by the temporal cortex, and also the entorhinal cortex and its connections with the amygdala and hippocampus (Mishkin 1978, 1980). Cells in the temporal cortex of macaques respond specifically to complex visual stimuli such as faces. Certain cells respond preferentially to changes in orientation of the face or changes in facial expression (Perrett *et al.* 1985). Facial communication plays an important role in recognition among primates and in the co-ordination of sexual activity in both sexes. Thus 'eye-contact proceptivity' is an important prelude to copulation in the chacma baboon (Bielert 1986). The proceptive displays of female marmosets, described earlier, are frequently stimulated by eye-eye contact with male partners (A. F. Dixson, unpublished observations). It is possible that visual stimuli of relevance to sexual arousal in the male, or proceptivity in the female, reach the hypothalamus via connections between the temporal cortex and limbic system (entorhinal cortex, amygdala, hippocampus).

Neurotransmitters, the hypothalamus and sexual behaviour in males

Numerous studies, mainly in rodents, have implicated monoamines in the control of masculine copulatory behaviour (Meyerson, Malmnas & Everitt 1985). Serotonin (5-HT) is thought to exert inhibitory influences upon copulatory behaviour whereas the catecholamines (dopamine and noradrenaline) have the opposite effect (McIntosh & Barfield 1984a,b,c). The MFB contains axons of catecholaminergic and 5-HT neurones with cell bodies in the mid- and hind-brain. Thus MFB lesions, which cause decreases in mounting behaviour, damage these ascending aminergic fibre systems as well as outputs descending from the POA. Destruction of cell bodies in the POA of male rats, with ibotenic acid, causes a decline in mounting and an inability to execute the deep pelvic thrust required for intromission; administration of lisuride restores intromission and ejaculation (Hansen *et al.* 1982). Lisuride, which also affects 5-HT neurones, is a dopamine agonist, and stimulation of dopaminergic inputs to the POA may activate the damaged circuitry that controls copulatory behaviour. In men suffering from Parkinson's disease, treatment with l-dopa sometimes enhances sexual activity (Bowers, Van Woert & Davis 1971). Dopaminergic mechanisms are also important in relation to sexual activity because dopamine acts as a prolactin-inhibiting factor and hyperprolactinaemia is associated with 'impotence' in man (Perryman & Thorner 1981) and copulatory deficiencies in male rats (Bailey & Herbert 1982). It has been suggested that chronic elevation of circulating prolactin impairs LHRH or dopaminergic influences upon sexual behaviour (e.g. Doherty, Bartke & Smith 1981). However, it is noteworthy that the effects of hyperprolactinaemia upon sexual behaviour develop gradually (e.g. Bailey & Herbert 1982) and resemble those seen in testosterone deficiency (e.g. in man: Bancroft, O'Carroll, McNeilly & Shaw 1984). Electrophysiological studies of the male rat have shown that refractory periods of some neurones projecting from the amygdala to the POA in the stria terminalis increase during chronic hyperprolactinaemia. The refractory periods of these neurones resemble those of castrated males, but normal concentrations of circulating testosterone can be maintained by implanting silastic capsules of hormone subcutaneously into hyperprolactinaemic males (Kendrick & Dixson 1984b). Thus chronic hyperprolactinaemia may impair sexual behaviour by decreasing central sensitivity to testosterone.

There is some evidence that the adrenergic system may be concerned in the control of masculine sexual arousal. Yohimbine, which is a selective antagonist of α_2-adrenoceptors (Goldberg & Robertson 1983), has been used in the treatment of human impotence (e.g. Morales, Surridge & Marshall 1981). Clark, Smith & Davidson (1985) found that in male rats yohimbine enhanced sexual activity by reducing the ejaculatory latency and post-ejaculatory interval. The same authors also reported that blockade of α_1 receptors by prazosin decreased sexual activity in male rats, as did treatment with clonidine, a hypertensive drug which stimulates central α adrenoceptors. These studies indicate that adrenergic mechanisms may influence masculine sexual arousal by blockade of α_2 or stimulation of α_1 receptor mechanisms. Further studies in non-human primates would be valuable, because sexual dysfunction as a side-effect of hypertensive drugs occurs in man (e.g. Laver 1974).

Charney & Heninger (1986) found that yohimbine alone (30 mg, orally) had no effect upon penile erection in healthy male volunteers. However, when combined with naloxone (1 mg/kg, intravenously) it caused erections lasting for at least one hour. The opioid antagonist naloxone has been reported to stimulate copulatory behaviour in experienced or sexually sluggish male rats (Myers & Baum 1979). However, attempts to stimulate sexual activity in male monkeys with naloxone (rhesus: Abbott, Holman, Berman, Neff & Goy 1984) or naltrexone (talapoin: Meller, Keverne & Herbert 1980) were unsuccessful. In the rat there is good evidence that β-endorphin suppresses copulatory activity when infused into the lateral ventricles (McIntosh, Vallano & Barfield 1980) or into the POA (Hughes, Everitt & Herbert 1987). In the latter studies, naloxone blocked the suppressive effects of centrally administered β-endorphin upon copulatory behaviour.

Clearly much remains to be learned of the effects of opioids and other neuroactive peptides, such as LHRH, upon sexual behaviour in male primates. In castrated rats LHRH increases sexual arousal, as indicated by reduced mount and ejaculatory latencies, provided that some testosterone is present (Moss 1978). Several studies have implicated LHRH in facilitation of sexual arousal in normal or impotent men (Evans & Distiller 1979; Benkert *et al.* 1975). Studies in non-human primates, to examine the effects of intracranial rather than peripheral administration of LHRH or its antagonists, would also be valuable.

Discussion

Although the hormonal regulation of sexual receptivity may be less rigid in female monkeys or apes than in many other mammals, there are numerous examples of important actions of ovarian hormones upon primate sexual behaviour. The effects of oestrogen and progesterone upon sexual attractiveness in species such as the rhesus monkey or baboon make it difficult to decide whether these hormones also act upon the brain to alter proceptivity or receptivity during the ovarian cycle. Experiments with hormonal implants in the brain are required to resolve these questions. Such studies indicate that testosterone propionate influences proceptivity and receptivity in female rhesus monkeys (Everitt & Herbert 1975). Oestradiol 17β stimulates proceptivity and to a lesser degree sexual receptivity in ovariectomized marmosets. Progesterone has inhibitory effects on both these aspects of sexual behaviour in the marmoset, at least when behaviour is observed during paired encounters. The opportunity therefore exists to examine the central effects of oestradiol or progesterone by implanting them into the hypothalamus of marmosets.

The lesion studies reported in this review indicate that gross damage of the anterior and medial hypothalamus causes a striking decrease in proceptivity in female marmosets, whereas sexual receptivity is usually unaffected. The role of hypothalamic neurones, or of fibre systems that traverse the hypothalamus, in the control of proceptivity and receptivity may therefore differ in important ways. Since female anthropoid primates do not use reflexive lordotic postures during mating, the neural substrates of receptivity in rats (Pfaff 1980) may not be paralleled in the hypothalamus of monkeys and apes. However, it remains to be seen whether lesions of the ventromedial hypothalamus, which disrupt lordosis in rodents, have any effect upon sexual receptivity in anthropoids. Since prosimian primates exhibit restricted periods of sexual receptivity or 'oestrus' there is a strong supposition that the neuroendocrine control of their sexual behaviour is akin to that seen in rodents and other mammals. Unfortunately, information on the effects of ovariectomy or hormone replacement in prosimians is scanty and there have been no studies of hypothalamic mechanisms in this group.

The capacity of some primate species to mate throughout the ovarian cycle, or after ovariectomy, may be greater when they are observed in the laboratory than under more natural conditions. However, it would be wrong to dismiss all such effects as the result of 'abnormal' environmental conditions. Laboratory experiments in monkeys or apes accentuate a propensity which exists (e.g. in the marmoset, rhesus monkey or orang-utan) to free sexual receptivity from control by oestrogen and progesterone during the peri-ovulatory phase of the cycle. This is an ancient phenomenon since it occurs in callitrichids and owl monkeys (Dixson 1983c) of the New World, as well as in some Old World monkeys and apes. Human ancestors also inherited this characteristic. Some anthropologists make a major distinction between human sexuality and the occurrence of 'oestrus' in other primates, but the experimental data do not support this view.

Given the complexity of social organization in monkeys and apes it was understandable that Zuckerman (1932) should propose that sexual behaviour constitutes a binding force in primate social life. However, explanations of the 'loss of oestrus' during primate evolution based on proximate mechanisms are not convincing, particularly given modern knowledge of primate reproduction gleaned from numerous field studies. A detailed analysis of mating strategies in free-ranging primates, coupled with studies of hormone-behaviour relationships in the laboratory is more likely to yield clues about the evolution of sexual behaviour in the order as a whole. There may be selective advantages for mechanisms that allow a greater degree of 'female choice' in sexual encounters (Taub 1980; Milton 1985; Dixson 1987b). Hormonal stimulation of female proceptivity and attractiveness may be especially important in this context. The capacity to 'choose', either actively or passively, between males may also increase if periods of sexual receptivity are extended and freed from absolute control by ovarian hormones.

In male primates it appears that testicular hormones and hypothalamic mechanisms play a vital role in sexual arousal and copulatory behaviour, as in other vertebrates. However, there is now evidence from electrophysiological studies of macaques that a 'dual mechanism' may exist in the hypothalamus, such that the POA has an important role in integrating sexual arousal and the dorsomedial or posterior hypothalamus controls aspects of the copulatory mechanism (Perachio *et al.* 1979; Yoshimatsu 1983; Oomura *et al.* 1984). Further studies of the dorsomedial and posterior hypothalamus as related to sexual behaviour in primates would be most valuable. As yet we know relatively little about how sexual stimuli impinge upon hypothalamic mechanisms in primates. Some attention has been given to olfactory and tactile cues from the female which influence sexual behaviour in male monkeys. However, visual communication is of tremendous importance in primate sexual behaviour, in the control both of masculine sexual arousal and of proceptivity in females. Experiments on memory (Mishkin 1978) and the physiology of facial recognition in macaques (Perrett *et al.* 1985) have shown that the temporal cortex, entorhinal cortex, amygdala and hippocampus are important in the complex processing of visual inputs that affect learning and emotional responses. The relevance of these findings to studies of the hypothalamus and primate sexual behaviour remains to be determined.

Acknowledgements

The research on marmosets described here has received generous financial support from the Medical Research Council and the Wellcome Trust (UK).

References

Abbott, D.H., Holman, S.D., Berman, M., Neff, D.A. & Goy, R.W. (1984). Effects of opiate antagonists on hormones and behavior of male and female rhesus monkeys. *Archs sex. Behav.* **13**: 1–25.

Bailey, D.J. & Herbert, J. (1982). Impaired copulatory behavior of male rats with hyperprolactinaemia induced by domperidone or pituitary grafts. *Neuroendocrinology* **35**: 186–93.

Bancroft, J. (1983). *Human sexuality and its problems.* Churchill Livingstone, Edinburgh.

Bancroft, J., O'Carroll, R., McNeilly, A. & Shaw, R.W. (1984). The effects of bromocriptine on the sexual behaviour of hyperprolactinaemic man: a controlled case study. *Clin. Endocr.* **21**: 131–7.

Barfield, R.J. (1969). Activation of copulatory behaviour by androgen implanted into the preoptic area of the male fowl. *Horm. Behav.* **1**: 37–51.

Barfield, R.J., Rubin, B.S., Glaser, J.H. & Davis, P.G. (1983). Sites of action of ovarian hormones in the regulation of oestrous responsiveness in rats. In *Hormones and behavior in higher vertebrates*: 2–17. (Eds Balthazart, J., Prove, E. & Gilles, R.). Springer, Berlin.

Barry, J. (1976). Characterization and topography of LHRH neurons in the human brain. *Neurosci. Lett.* **3**: 287–91.

Barry, J. (1978). Immunofluorescence study of the preoptico-terminal LRH tract in the female squirrel monkey during the estrous cycle. *Cell Tissue Res.* **198**: 1–13.

Baum, M.J., Keverne, E.B., Everitt, B.J. & Herbert, J. (1976). Reduction of sexual interaction in rhesus monkeys by a vaginal action of progesterone. *Nature, Lond.* **263**: 606–8.

Baum, M.J., Slob, A.K., de Jong, F.H. & Westbroek, D.L. (1978). Persistence of sexual behaviour in ovariectomized stumptail macaques following dexamethasone treatment or adrenalectomy. *Horm. Behav.* **11**: 323–47.

Beach, F.A. (1956). Characteristics of masculine "sex drive". In *The Nebraska symposium on motivation* **4**: 1–31. (Ed. Jones, M.R.). Univ. Nebraska Press, Lincoln.

Beach, F.A. (1970). Coital behaviour in dogs. VI. Long-term effects of castration upon mating in the male. *J. comp. physiol. Psychol.* **70** (3) Part 2, *Monogr.*: 1–32.

Beach, F.A. (1976). Sexual attractivity, proceptivity, and receptivity in female mammals. *Horm. Behav.* **7**: 105–38.

Bean, N.J., Nunez, A.A. & Conner, R. (1981). Effects of medial preoptic lesions on male mouse ultrasonic vocalizations and copulatory behavior. *Brain. Res. Bull.* **6**: 109–12.

Benkert, O., Jordan, R., Dahlen, H.G., Schneider, H.P.G. & Gammel, G. (1975). Sexual impotence; a double blind study of LHRH nasal spray versus placebo. *Neuropsychobiology* **1**: 203–10.

Bielert, C. (1986). Sexual interactions between captive adult male and female Chacma baboons (*Papio ursinus*) as related to the female's menstrual cycle. *J. Zool., Lond.* **209**: 521–36.

Bielert, C., Howard-Tripp, M.E. & Van Der Walt, L.A. (1980). Environmental and social factors influencing seminal emission in chacma baboons (*Papio ursinus*). *Psychoneuroendocrinology* **5**: 287–303.

Bonney, R.C., Dixson, A.F. & Fleming, D. (1980). Plasma concentrations of oestradiol-17β, oestrone, progesterone and testosterone during the ovarian cycle of the owl monkey (*Aotus trivirgatus*). *J. Reprod. Fert.* **60**: 101–7.

Bonsall, R.W., Rees, H.D. & Michael, R.P. (1986). [^{3}H] Estradiol and its metabolites in the brain, pituitary gland and reproductive tract of the male rhesus monkey. *Neuroendocrinology* **43**: 98–109.

Bowers, M.B., van Woert, M. & Davis, L. (1971). Sexual behavior during L-dopa treatment for Parkinsonism. *Am. J. Psychiat.* **127**: 1691.

Chan, A., Dudley, C.A. & Moss, R.L. (1984). Hormonal and chemical modulation of ventromedial hypothalamic neurons responsive to vagino-cervical stimulation. *Neuroendocrinology* **38**: 328–36.

Charney, D.S. & Heninger, G.R. (1986). α-2-Adrenergic and opiate receptor blockade: synergistic effects on anxiety in healthy subjects. *Archs gen. Psychiatry* **43**: 1037–41.

Clark, J.T., Smith, E.R. & Davidson, J.M. (1985). Evidence for modulation of sexual behaviour by α-adrenoceptors in male rats. *Neuroendocrinology* **41**: 36–43.

Clegg, M.T., Santolucito, J.A., Smith, J.D. & Ganong, W.F. (1958). The effect of hypothalamic lesions on sexual behavior and estrous cycles in the ewe. *Endocrinology* **62**: 790–7.

Crews, D. & Silver, R. (1985). Reproductive physiology and behavior interactions in nonmammalian vertebrates. In *Handbook of behavioral neurobiology* 7: 101–82. (Eds Adler, N., Pfaff, D. & Goy, R.W.). Plenum, New York.

Czaja, J.A. & Bielert, C. (1975). Female rhesus sexual behavior and distance to male partner: relation to stage of the menstrual cycle. *Archs sex. Behav.* **4**; 583–97.

Davidson, J.M. (1966). Activation of the male rat's sexual behavior by intracerebral implantation of androgen. *Endocrinology* **79**: 783–94.

Demski, L.S., Bauer, D.H. & Gerald, J.W. (1975). Sperm release evoked by electrical stimulation of the fish brain: a functional-anatomical study. *J. exp. Zool.* **191**: 215–32.

Dixson, A.F. (1983a). The hormonal control of sexual behaviour in primates. *Oxford Revs reprod. Biol.* **5**: 131–219.

Dixson, A.F. (1983b). Observations on the evolution and behavioral significance of 'sexual skin' in female primates. *Adv. Study Behav.* **13**: 63–106.

Dixson, A.F. (1983c). The owl monkey (*Aotus trivirgatus*). In *Reproduction in New World primates*: 69–113. (Ed. Hearn, J.P.). MTP Press, Lancaster, Boston & The Hague.

Dixson, A.F. (1986a). Proceptive displays of the female common marmoset (*Callithrix jacchus*): effects of ovariectomy and oestradiol 17β. *Physiol. Behav.* **36**: 971–3.

Dixson, A.F. (1986b). Genital sensory feedback and sexual behaviour in male and female marmosets (*Callithrix jacchus*). *Physiol. Behav.* **37**: 447–50.

Dixson, A.F. (1987a). Effects of adrenalectomy upon proceptivity, receptivity and sexual attractiveness in ovariectomized common marmosets (*Callithrix jacchus*). *Physiol. Behav.* **39**: 495–99.

Dixson, A.F. (1987b). Observations on the evolution of the genitalia and copulatory behaviour in male primates. *J. Zool., Lond.* **213**: 423–43.

Dixson, A.F., Everitt, B.J., Herbert, J., Rugman, S.M. & Scruton, D. (1973). Hormonal and other determinants of sexual attractiveness and receptivity in rhesus and talapoin monkeys. In *Symposia of the fourth international congress of primatology, Portland, Ore., August 15–18 1972. 2. Primate reproductive behaviour*: 36–63. (Ed. Phoenix, C.H.). Karger, Basel.

Dixson, A.F. & Herbert, J. (1977). Gonadal hormones and sexual behavior in groups of adult talapoin monkeys (*Miopithecus talapoin*). *Horm. Behav.* **8**: 141–54.

Dixson, A.F. & Lunn, S.F. (1987). Post-partum changes in hormones and sexual behaviour in captive groups of marmosets (*Callithrix jacchus*). *Physiol. Behav.* **41**: 577–83.

Doherty, P.C., Bartke, A. & Smith, M.S. (1981). Differential effects of bromocriptine treatment on LH release and copulatory behavior in hyperprolactinaemic male rats. *Horm. Behav.* **15**: 436–50.

Eaton, G.G. (1973). Social and endocrine determinants of sexual behavior in simian and prosimian females. In *Symposia of the fourth international congress of primatology, Portland, Ore., August 15–18 1972. 2. Primate reproductive behaviour*: 20–35. (Ed. Phoenix, C.H.). Karger, Basel.

Edwards, D.A. & Matthews, D. (1977). The ventromedial nucleus of the hypothalamus and the hormonal arousal of sexual behaviors in the female rat. *Physiol. Behav.* **14**: 439–53.

Evans, I.M. & Distiller, L.A. (1979). Effects of luteinizing hormone releasing hormone on sexual arousal in normal men. *Archs sex. Behav.* **8**: 385–95.

Everitt, B.J. & Herbert, J. (1971). The effects of dexamethasone and androgens on sexual receptivity of female rhesus monkeys. *J. Endocr.* **51**: 575–88.

Everitt, B.J. & Herbert, J. (1972). Hormonal correlates of sexual behavior in subhuman primates. *Dan. med. Bull.* **19**: 246–55.

Everitt, B.J. & Herbert, J. (1975). The effects of implanting testosterone propionate into the central nervous system on the sexual behavior of adrenalectomized female rhesus monkeys. *Brain Res.* **86**: 10–20.

Everitt, B.J., Herbert, J. & Hamer, J.D. (1972). Sexual behavior of bilaterally adrenalectomized female rhesus monkeys. *Physiol. Behav.* **8**: 409–15.

Feder, H.H. (1978). Regulation of sexual behavior by hormones in female nonprimates. In *Handbook of sexology* **2**: 393–411. (Eds Money, J. & Mustaph, H.). Elsevier, New York.

Gillman, J. & Stein, H.B. (1941). A quantitative study of the inhibition of oestradiol benzoate by progesterone in the baboon (*Papio porcarius*). *Endocrinology* **28**: 274–82.

Girolami, L. & Bielert, C. (In press). Female perineal sexual swelling: a sexual releaser in the chacma baboon (*Papio ursinus*). *Int. J. Primatol.*

Goldberg, M.R. & Robertson, D. (1983). Yohimbine: a pharmacological probe for the study of the α_2 adrenoceptor. *Pharm. Revs* **35**: 143–80.

Goy, R.W. & Resko, J.A. (1972). Gonadal hormones and behavior of normal and pseudohermaphroditic nonhuman female primates. *Rec. Prog. Horm. Res.* **28**: 707–33.

Grant, E.C.G. & Pryse Davies, J. (1968). Effect of oral contraception on depressive mood changes and on endometrial monoamine oxidase and phosphatases. *Br. med. J.* **1968** (3): 777–80.

Green, J.D., Clemente, C.D. & De Groot, J. (1957). Rhinencephalic lesions and behavior in cats. *J. comp. Neurol.* **108**: 505–45.

Hansen, S. (1982). Hypothalamic control of motivation: the medial preoptic area and masculine sexual behaviour. *Scand. J. Psychol., Suppl.* **1**: 121–6.

Hansen, S., Kohler, Ch., Goldstein, M. & Steinbusch, H.V.M. (1982). Effects of ibotenic acid-induced neuronal degeneration in the medial preoptic area and the lateral hypothalamic area on sexual behavior in the male rat. *Brain Res.* **239**: 213–32.

Harcourt, A.H., Fossey, D., Stewart, K.J. & Watts, D.P. (1980). Reproduction in wild gorillas and some comparisons with chimpanzees. *J. Reprod. Fert., Suppl.* **28**: 59–70.

Harris, V.H. & Sachs, B.D. (1975). Copulatory behavior in male rats following amygdaloid lesions. *Brain. Res* **86**: 514–8.

Hart, B.L. (1974). The medial preoptic-anterior hypothalamic area and sociosexual behavior of male dogs: a comparative neuropsychological analysis. *J. comp. physiol. Psychol.* **86**: 328–9.

Hart, B.L. (1986). Medial preoptic-anterior hypothalamic lesions and sociosexual behavior of male goats. *Physiol. Behav.* **36**: 301–5.

Hart, B.L., Haugen, C.M. & Peterson, D.M. (1973). Effects of medial preoptic-anterior hypothalamic lesions on mating behavior of male cats. *Brain. Res.* **54** 177–91.

Hart, B.L. & Jones, T.O.A.C. (1975). Effects of castration on sexual behavior of tropical male goats. *Horm. Behav.* **6**: 247–58.

Hart, B.L. & Leedy, M.G. (1985). Neurological bases of male sexual behavior. In *Handbook of behavioral neurobiology* **7**: 373–422. (Eds Adler, N., Pfaff, D. & Goy, R.W.). Plenum, New York.

Heimer, L. & Larsson, K. (1967). Impairment of mating behavior in male rats following lesions in the preoptic-anterior hypothalamic continuum. *Brain. Res.* **3**: 248–63.

Herbert, J. (1966). The effect of oestrogen applied directly to the genitalia upon the sexual attractiveness of the female rhesus monkey. *Excerpta med. int. Congr. Ser.* **3**: 212.

Herbert, J. (1970). Hormones and reproductive behavior in rhesus and talapoin monkeys. *J. Reprod. Fert.* **11**: 119–40.

Herbert, J. (1973). The role of the dorsal nerves of the penis in the sexual behaviour of the male rhesus monkey. *Physiol. Behav.* **10**: 293–300.

Herbert, J. (1974). Some functions of hormones and the hypothalamus in the sexual activity of primates. *Prog. Brain Res.* **41**: 331–48.

Herbert, J. (1978). The neuroendocrine basis of sexual behavior in primates. In *Handbook of sexology* **2**: 449–59. (Eds Money, J. & Mustaph, H.). Elsevier, New York.

Herbert, J. & Trimble, M. (1967). Effect of oestradiol and testosterone on the sexual receptivity and attractiveness of the female rhesus monkey. *Nature, Lond.* **216**: 165–6.

Hess, D.L. & Resko, K.A. (1973). The effects of progesterone on the patterns of testosterone and oestradiol concentrations in the systemic plasma of the female rhesus monkey during the intermenstrual period. *Endocrinology* **92**: 446–53.

Hughes, A.M., Everitt, B.J. & Herbert, J. (1987). Selective effects of β-endorphin infused into the hypothalamus, preoptic area and bed nucleus of the stria terminalis on the sexual and ingestive behaviour of male rats. *Neuroscience* **23**: 1063–73.

Hutchison, J.B. (1978). Hypothalamic regulation of male sexual responsiveness to androgen. In *Biological determinants of sexual behaviour*: 277–317. (Ed. Hutchison, J.B.). Wiley, Chichester.

Janson, C.H. (1984). Female choice and mating system of the brown capuchin monkey (*Cebus apella*). *Z. Tierpsychol.* **65**: 177–200.

Johnson, D.F. & Phoenix, C.H. (1976). Hormonal control of female sexual attractiveness, proceptivity and receptivity in rhesus monkeys. *J. comp. physiol. Psychol.* **90**: 473–83.

Kalra, S.P. & Kalra, P.S. (1983). Neural regulation of luteinizing hormone secretion in the rat. *Endocr. Revs* **4:** 311–51.

Kendrick, K.M. (1983). Electrophysiological effects of testosterone on the medial preoptic anterior hypothalamus of the rat. *J. Endocr.* **96:** 35–42.

Kendrick, K.M. & Dixson, A.F. (1983). The effect of the ovarian cycle on the sexual behaviour of the common marmoset (*Callithrix jacchus*). *Physiol, Behav.* **30:** 735–42.

Kendrick, K.M. & Dixson, A.F. (1984a). Ovariectomy does not abolish proceptive behaviour cyclicity in the common marmoset (*Callithrix jacchus*). *J. Endocr.* **101:** 155–62.

Kendrick, K.M. & Dixson, A.F. (1984b). Hyperprolactinaemia abolishes sensitivity of stria terminalis neurones to testosterone. *J. Endocr.* **102:** 109–13.

Kendrick, K.M. & Dixson, A.F. (1985a). Effects of oestradiol 17β, progesterone and testosterone upon proceptivity and receptivity in ovariectomized common marmosets (*Callithrix jacchus*). *Physiol. Behav.* **34:** 123–8.

Kendrick, K.M. & Dixson, A.F. (1985b). Luteinizing hormone releasing hormone enhances proceptivity in a primate. *Neuroendocrinology* **41:** 449–53.

Kendrick, K.M. & Dixson, A.F. (1986). Anteromedial hypothalamic lesions block proceptivity but not receptivity in the female common marmoset (*Callithrix jacchus*). *Brain Res.* **375:** 221–9.

Kendrick, K.M., Drewett, R.F. & Wilson, C.A. (1981). Effect of testosterone on neuronal refractory periods, sexual behaviour and luteinizing hormone: a comparison of time courses. *J. Endocr.* **89:** 147–55.

Kluver, N. & Bucy, P.C. (1939). Preliminary analysis of functions of temporal lobes in monkeys. *Archs Neurol. Psychiat.* **42:** 979–1000.

Komisaruk, B.R. (1978). The nature of the neural substrate of female sexual behaviour in mammals and its hormonal sensitivity: review and speculations. In *Biological determinants of sexual behaviour:* 349–93. (Ed. Hutchison, J.B.). Wiley, Chichester.

Larsson, K. (1979), Features of the neuroendocrine regulation of masculine sexual behavior. In *Endocrine control of sexual behavior*: 77–163. (Ed. Beyer, C.). Raven Press, New York.

Laver, M.C. (1974). Sexual behavior patterns in male hypertensives. *Aust. N.Z.J. Med.* **4:** 29–31.

Law, T. & Meagher, W. (1958). Hypothalamic lesions and sexual behavior in the female rat. *Science, N.Y.* **128:** 1626–7.

Lipp, H.P. & Hunsperger, R.W. (1978). Threat, attack and flight elicited by electrical stimulation of the ventromedial hypothalamus of the marmoset monkey *Callithrix jacchus. Brain, Behav. Evolut.* **15:** 260–93.

McIntosh, T.K. & Barfield, R.J. (1984a). Brain monoaminergic control of male reproductive behavior. I. Serotonin and the post-ejaculatory refractory period. *Behav. Brain. Res.* **12:** 255–65.

McIntosh, T.K. & Barfield, R.J. (1984b). Brain monoaminergic control of male reproductive behavior. II. Dopamine and the post-ejaculatory refractory period. *Behav. Brain. Res.* **12:** 267–73.

McIntosh, T.K. & Barfield, R.J. (1984c). Brain monoaminergic control of male reproductive behavior. III. Norepinephrine and the post-ejaculatory refractory period. *Behav. Brain Res.* **12:** 275–81.

McIntosh, T.K., Vallano, M.L. & Barfield, R.J. (1980). The effects of morphine, B-endorphin and naloxone on catecholamine levels and sexual behavior in the male rat. *Pharmacol. Biochem. Behav.* **13:** 435–41.

MacLean, P.D. & Ploog, D.W. (1962). Cerebral representation of penile erection. *J. Neurophysiol.* **25:** 29–55.

Malsbury, C., Kow, L.M. & Pfaff, D.W. (1977). Effects of medial hypothalamic lesions on lordosis and other behaviors in female hamsters. *Physiol. Behav.* **19:** 223–37.

Mangat, H.K., China, G.S., Singh, B. & Anand, B.K. (1978). Influence of gonadal hormones and genital efferents on E.E.G. activity of the hypothalamus in adult male rhesus monkeys. *Physiol. Behav.* **20:** 377–84.

Marshall, P.E. & Goldsmith, P.C. (1980). Neuroregulatory and neuroendocrine GnRH pathways in the hypothalamus and forebrain of the baboon. *Brain. Res.* **193:** 353–72.

Martin, G.B. (1984). Factors affecting the secretion of luteinizing hormone in the ewe. *Biol. Rev.* **59:** 1–87.

Meller, R.E., Keverne, E.B. & Herbert, J. (1980). Behavioural and endocrine effects of naltrexone in male talapoin monkeys. *Pharmacol. Biochem. Behav.* **13:** 663–72.

Merari, A. & Ginton, A. (1975). Characteristics of exaggerated sexual behavior induced by electrical stimulation of the medial preoptic area in male rats. *Brain Res.* **86:** 97–108.

Meyers, R. (1961). Evidence of a locus in the neural mechanisms of libido and penile potency in the septo-fornico-hypothalamic region of the human brain. *Trans. Am. neurol. Ass.* **86:** 81–5.

Meyers, R. (1962). Three cases of myoclonus alleviated by bilateral ansotomy, with a note on postoperative albido and impotence. *J. Neurosurgery* **19:** 71–81.

Meyerson, B., Malmnas, C.O. & Everitt, B.J. (1985). Neuropharmacology, neurotransmitters and sexual behavior in mammals. In *Handbook of behavioral neurobiology* **7:** 495–536. (Eds Adler, N., Pfaff, D. & Goy, R.W.). Plenum, New York.

Michael, R.P. (1969). Neural and non-neural mechanisms in the reproductive behavior of primates. In *Progress in endocrinology*: 302–9. (Ed. Gjal, C.). Excerpta Medica, Amsterdam.

Michael, R.P. (1971). Neuroendocrine factors regulating primate behavior. In *Frontiers in neuroendocrinology*: 359–98. (Eds Martini, L. & Ganong, W.F.). Oxford University Press, London.

Michael, R.P. & Bonsall, R.W. (1979). Hormones and the sexual behavior of rhesus monkeys. In *Endocrine control of sexual behavior*: 279–302. (Ed. Beyer, C.). Raven Press, New York.

Michael, R.P. & Keverne, E.B. (1968). Pheromones in the communication of sexual status in primates. *Nature, Lond.* **218:** 746–9.

Michael, R.P., Richter, M.C., Cain, J.A., Zumpe, D. & Bonsall, R.W. (1978). Artificial menstrual cycles, behaviour and the role of androgens in female rhesus monkeys. *Nature, Lond.* **275:** 439–40.

Michael, R.P., Saayman, G.S. & Zumpe, D. (1968). The suppression of mounting behaviour and ejaculation in male rhesus monkeys (*Macaca mulatta*) by administration of progesterone to their female partners. *J. Endocr.* **41:** 421–31.

Michael, R.P. & Wilson, M. (1974). Effects of castration and hormone replacement in fully adult male rhesus monkeys (*Macaca mulatta*). *Endocrinology* **95:** 150–9.

Michael, R.P. & Zumpe, D. (1977). Effects of androgen administration on sexual invitations by female rhesus monkeys (*Macaca mulatta*). *Anim. Behav.* **25:** 936–44.

Milton, K. (1985). Mating patterns of woolly spider monkeys, *Brachyteles arachnoides*: implications for female choice. *Behav. Ecol. Sociobiol.* **17:** 53–9.

Mishkin, M. (1978). Memory in monkeys severely impaired by combined but not by separate removal of amygdala and hippocampus. *Nature, Lond.* **273:** 297–8.

Mishkin, M. (1980). A memory system in the monkey. *Phil. Trans. R. Soc.* (B) **298:** 85–95.

Morales, A. Surridge, D.H. & Marshall, P.G. (1981). Yohimbine for treatment of impotence in diabetics. *N. Engl. J. Med.* **305:** 1221.

Morali, G. & Beyer, C. (1979). Neuroendocrine control of mammalian estrous behavior. In *Endocrine control of sexual behavior*: 33–75. (Ed. Beyer, C.). Raven Press, New York.

Moss, R.L. (1978). Effects of hypothalamic peptides on sex behavior in animals and man. In *Psychopharmacology: A generation of progress*: 431–40. (Eds Lipton, M.A., DiMascio, A. & Killam, K.F.). Raven Press, New York.

Moss, R.L. & Foreman, M.M. (1976). Potentiation of lordosis behavior by intrahypothalamic infusion of synthetic luteinizing hormone releasing hormone. *Neuroendocrinology* **20:** 176–81.

Moss, R.L. & McCann, S.M. (1973). Induction of mating behavior in rats by luteinizing hormone-releasing factor. *Science, N.Y.* **181:** 177–9.

Myers, B.M. & Baum, M.J. (1979). Facilitation by opiate antagonists of sexual performance in the male rat. *Pharmacol. Biochem. Behav.* **10:** 615–18.

Nadler, R.D. (1982). Laboratory research on sexual behaviour and reproduction of gorillas and orang-utans. *Am. J. Primatol. Suppl.* No. 1: 57–66.

Nadler, R.D., Graham, C.E., Collins, D.C. & Gould, K.G. (1979). Plasma gonadotropins, prolactin, gonadal steroids and genital swelling during the menstrual cycle of lowland gorillas. *Endocrinology* **105**: 290–6.

Nance, D.M., Shryne, J. & Gorski, R.A. (1975). Effects of septal lesions on behavioural sensitivity of female rats to gonadal hormones. *Horm. Behav.* **6**: 59–64.

Oomura, Y., Yoshimatsu, H. & Auo, S. (1984). Hypothalamic neuronal activity during sexual behaviour in male monkey. In *Modulation of sensorimotor activity during alterations in behavioral states*: 231–50. (Ed. Bandler, R.). Alan R. Riss Inc., New York.

Paxinos, G. & Bindra, D. (1971). Hypothalamic knife cuts: effects on eating, drinking, irritability, aggression and copulation in the male rat. *J. comp. physiol. Psychol.* **79**: 219–29.

Perachio, A.A., Marr, L.D. & Alexander, M. (1979). Sexual behaviour in male rhesus monkeys elicited by electrical stimulation of preoptic and hypothalamic areas. *Brain Res.* **177**: 127–44.

Perrett, D.I., Smith, P.A.J., Potter, D.D., Mistlin, A.J., Head, A.S., Milner, A.D. & Jeeves, M.A. (1985). Visual cells in the temporal cortex sensitive to face view and gaze direction. *Proc. R. Soc.* (B) **223**: 293–317.

Perryman, R.L. & Thorner, M.O. (1981). The effects of hyperprolactinaemia on sexual and reproductive function in man. *J. Androl.* **2**: 233–42.

Pfaff, D.W. (1980). *Estrogens and brain function.* Springer, New York.

Pfaff, D. & Modianos, D. (1985). Neural mechanisms of female reproductive behavior. In *Handbook of behavioral neurobiology* 7: 423–93. (Eds Adler, N., Pfaff, D. & Goy, R.W.). Plenum, New York.

Pfaff, D.W. & Pfaffman, C. (1969). Olfactory and hormonal influences on the basal forebrain of the male rat. *Brain Res.* **75**: 137–56.

Phoenix, C.H. (1973). The role of testosterone in the sexual behavior of laboratory male rhesus. In *Symposia of the fourth international congress of primatology, Portland, Ore., August 15–18 1972.* **2.** *Primate reproductive behaviour*: 99–122. (Ed. Phoenix, C.H.). Karger, Basel.

Phoenix, C.J. (1974). Effects of dihydrotesterone on sexual behavior of castrated male rhesus monkeys. *Physiol. Behav.* **12**: 1054–5.

Phoenix, C.H. & Chambers, K.C. (1982). Sexual behavior in adult gonadectomized female pseudohermaphrodite, female and male rhesus macaques (*Macaca mulatta*) treated with estradiol benzoate and testosterone propionate. *J. comp. physiol. Psychol.* **96**: 823–33.

Pohl, C.R. & Knobil, E. (1982). The role of the central nervous system in the control of ovarian function in higher primates. *A. Rev. Physiol.* **44**: 583–93.

Pomerantz, S.M. & Goy, R.W. (1983). Proceptive behavior of female rhesus monkeys during tests with tethered males. *Horm. Behav.* **17**: 237–48.

Powers, J.B. & Valenstein, E.S. (1972). Sexual receptivity: facilitation by medial preoptic lesions in female rats. *Science, N.Y.* **175**: 1003–5.

Rees, H.D. & Michael, R.P. (1982). Brain cells of the male rhesus monkey accumulate ^{3}H-testosterone or its metabolites. *J. comp. Neurol.* **206**: 273–7.

Robinson, B.M. & Mishkin, M. (1966). Ejaculation evoked by stimulation of the preoptic area in monkeys. *Physiol. Behav.* **1**: 269–72.

Saayman, G.S. (1973). Effects of ovarian hormones on the sexual skin and behavior of ovariectomized baboons (*Papio ursinus*) under free-ranging conditions. In *Symposia of the fourth international congress of primatology, Portland, Ore., August 15–18 1972.* **2.** *Primate reproductive behaviour:* 64–98. (Ed. Phoenix, C.H.). Karger, Basel.

Sakuma, Y. & Pfaff, D.W. (1980). LH-RH in the mesencephalic central grey can potentiate lordosis reflex of female rats. *Nature, Lond.* **283**: 566–7.

Sar, M. & Stumpf, W.E. (1973). Autoradiographic localization of radioactivity in the rat brain after injection of 1,2$-$H^3$-$testosterone. *Endocrinology* **92**: 251–6.

Schenck, P.E. & Slob, A.K. (1986). Castration, sex steroids and heterosexual behavior in adult, laboratory housed stumptailed macaques (*Macaca arctoides*). *Horm. Behav.* **20**: 336–53.

Schmidt, R.S. (1968). Preoptic activation of frog mating behavior. *Behaviour* **30**: 239–57.

Schürmann, C. (1982). Mating behaviour of wild orang utans. In *The orang utan: its biology and conservation:* 269–84. (Ed. de Boer, L.E.M.). Dr W. Junk, The Hague, Boston, London. (*Perspect. vert. Sci.* **2**).

Sheridan, P.J., Hagino, N. & Weaker, F.J. (1982). Androgen-concentrating cells in the periventricular brain of the female rhesus monkey. *J. comp. Neurol.* **207:** 93–8.

Sirinathsinghji, D.J.S. (1983). GnRH in the spinal subarachnoid space potentiates lordosis behavior in the female rat. *Physiol. Behav.* **33:** 717–23.

Slimp, J.C., Hart, B.L. & Goy, R.W. (1978). Heterosexual, autosexual and social behavior of adult male rhesus monkeys with medial preoptic-anterior hypothalamic lesions. *Brain Res.* **142:** 105–22.

Smith, E., Damassa, D.A. & Davidson, J.M. (1977). Plasma testosterone and sexual behavior following intracerebral implantation of testosterone propionate in the castrated male rat. *Horm. Behav.* **8:** 77–87.

Szechtman, H., Caggiula, A.R. & Wulkan, D. (1978). Preoptic knife cuts and sexual behavior in male rats. *Brain Res.* **150:** 569–91.

Taub, D.M. (1980). Female choice and mating strategies among wild Barbary macaques (*Macaca sylvanus* L.). In *The macaques: studies in ecology, behavior and evolution:* 287–344. (Ed. Lindburg, D.G.). Van Nostrand Reinhold, New York.

van Dis, H. & Larsson, K. (1970). Seminal discharge following intracranial electrical stimulation. *Brain Res.* **23:** 381–6.

Wada, M. & Gorbman, A. (1977a). Mate calling induced by electrical stimulation in freely moving leopard frogs (*Rana pipiens*). *Horm. Behav.* **9:** 141–9.

Wada, M. & Gorbman, A. (1977b). Relation of mode of administration of testosterone to evocation of male sex behavior in frogs. *Horm. Behav.* **8:** 310–9.

Wallen, K. (1982). Influence of female hormonal state upon rhesus monkey sexual behavior varies with space for social interaction. *Science, N.Y.* **217:** 375–7.

Wallen, K. & Goy, R.W. (1977). Effects of estradiol benzoate, oestrone and propionates of testosterone and dihydrotestosterone on sexual and related behaviors of ovariectomized rhesus monkeys. *Horm. Behav.* **9:** 228–48.

Yoshimatsu, H. (1983). Hypothalamic regulatory mechanism of sexual behavior in the male monkey. *Fukuoka Acta med.* **74:** 89–100.

Young, W.C. & Orbison, W.D. (1944). Changes in selected features of behavior in pairs of oppositely-sexed chimpanzees during the menstrual cycle and after ovariectomy. *J. comp. Psychol.* **37:** 107–43.

Zuckerman, S. (1932). *The social life of monkeys and apes.* Kegan Paul, Trench, Trubner, London.

Zumpe, D. & Michael, R.P. (1984). Low potency of intact male rhesus monkeys after long-term visual contact with their female partners. *Am. J. Primatol.* **6:** 241–52.

Symp. zool. Soc. Lond. (1988) No. 60 119–134

Embryo manipulation for the regulation of reproduction and disease

P. M. SUMMERS
and J. P. HEARN

MRC/AFRC Comparative Physiology Research Group
Institute of Zoology
The Zoological Society of London
Regent's Park
London NW1 4RY, UK

Synopsis

Embryo manipulation comprises a number of techniques including embryo transfer and the more complex procedures of micro-surgical division and embryo reconstruction, cryopreservation, and sexing of embryos. Embryo transfer is now a well established technique for agriculturally important animals, particularly in beef and dairy cattle, in which it makes possible a rapid increase in the number of offspring from genetically superior animals. These techniques have recently been applied to the breeding of exotic species, particularly ungulates, but progress has been restricted by deficiencies in knowledge of their reproductive physiology and by the need to use extra-specific surrogate mothers for endangered species.

Certain diseases can be controlled by embryo transfer. The intact zona pellucida is an effective barrier to bacteria and many viruses, and the embryo transfer carried out under appropriate conditions is less liable than the movement of live animals to transmit disease. The ability to freeze embryos of domestic animals simplifies their transport, but further work is needed on the cryopreservation of embryos from exotic species. However, it seems unlikely that animal health authorities will allow the importation of frozen embryos from animals in the wild.

Embryo manipulative procedures offer a means to study the basic mechanisms of innate resistance to specific diseases in certain cattle breeds. The production of *Bos indicus–Bos taurus* lymphoid chimeric twin calves and whole body chimeras provides a model for studying the roles of the immune system and other physiological and anatomical features in disease resistance.

The microsurgical division of embryos is commonly used in cattle to increase the number of embryos available for transfer and this technique will be useful for exotic species. Extra-specific embryo transfer is unsuccessful in some species combinations and this difficulty may be overcome by the microsurgical reconstruction of embryos.

Introduction

Almost a century has elapsed since the first experiments on manipulation of mammalian embryos were conducted by Walter Heape in Cambridge. In

ZOOLOGICAL SYMPOSIUM No. 60
ISBN 0-19-854007-8

those experiments he transferred rabbit embryos from one female to another. Up to about 1970, embryo manipulation remained a laboratory tool used to study the processes of reproduction, but since then tremendous technological advances have been made in the manipulation of pre-implantation embryos of mammals. The impetus has come from agriculture, particularly the beef and dairy cattle industries, in which embryo collection and transfer can increase the number of offspring from genetically desirable cows. Building on the knowledge obtained with animal embryos, *in vitro* fertilization and embryo transfer were rapidly developed for man and the technique is now a useful tool to alleviate certain forms of human infertility (Edwards 1985).

Embryo manipulation can take several forms, from the relatively simple procedures of collection of embryos from a donor animal and their transfer to suitably prepared recipients to the more complex procedures of micro-surgical division and reconstruction of embryos, cryopreservation, sexing of embryos and gene injection into the zygote. The possibility of using such technology for exotic species is an attractive one that has received recent consideration by reproductive physiologists (Durrant & Benirschke 1981; Stover & Westrom 1984). In this paper, we will review the procedures of embryo manipulation for the regulation of reproduction and disease in domestic animals and the possible applications in the breeding of captive and wild exotic species.

Embryo collection and transfer

Domestic animals

The collection of embryos and their transfer to recipient animals is now a well established technique in cattle, sheep, goats, pigs and horses (Allen 1982; Seidel & Seidel 1981). Underlying this success is the detailed knowledge of the reproductive physiology of these species which has accumulated from the efforts of research scientists over the past 60 years. In addition, successful embyro transfer is heavily dependent on a team of experienced personnel, from the stockmen responsible for oestrus detection in donors and recipients to the embryologist who accurately assesses embryo development and quality.

The technique is most widely applied to the domestic cow, both beef and dairy, in which it can be used to increase rapidly the number of calves born from a genetically superior cow. It has been predicted that in North America 200 000 pregnancies will be obtained by embryo transfer in 1988 (Seidel & Seidel 1981). In the cow, embryos are usually collected seven to eight days after oestrus by means of a non-surgical technique of flushing the uterus. The donor is treated to induce multiple ovulations and make possible the collection of normal embryos, preferably about eight; the procedure is highly variable, however, sometimes producing no embryos and sometimes as many as 30–40 of variable quality. Single embryos are transferred to the uterus of recipients whose oestrus was synchronous with that of the donor, by means of surgical or non-surgical transfer techniques. The former generally produce higher pregnancy rates of about 75%.

Similar techniques are used for the mare, but superovulation cannot be induced in this species. In the smaller species, embryo collection and transfer are usually done surgically.

Exotic species

Embryo transfer in exotic animals is in its infancy and its widespread application is limited by a number of factors. For example, knowledge of the reproductive physiology of exotic species, which underpins the successful outcome of the procedure, is often sparse. Unless such knowledge exists, or can be extrapolated from closely related domesticated species, embryo transfer is ruled out. Even some domesticated species require further study. The cat, for example, is attracting such study by those hoping to devise techniques applicable to the propagation of rare captive felids (Goodrowe, Howard & Wildt 1986).

A further difficulty arises because the technology is required mainly for endangered species, in which few if any females are available to act as recipients. As a result, surrogate mothers have to be selected from closely related domesticated or non-endangered exotic species.

Other limitations include the provision of adequate animal handling facilities, the need for personnel experienced in the collection and handling of embryos, and adequate financial support. Nevertheless, despite these difficulties, some progress has been made in recent years.

Embryo transfer in ungulates

It was shown by Bunch, Foote & Whitaker (1977) that mouflon (*Ovis musimon*) could be produced by embryo transfer to domestic sheep. A gaur calf (*Bos gaurus*) in 1981 in New York's Bronx Zoo was, however, the first example of a member of an endangered species born as a result of extra-specific embryo transfer to a surrogate mother, in this case a Holstein-Friesian cow (Stover, Evans & Dolensek 1981). Five blastocysts were collected from a superovulated gaur cow; of three recipients found to be pregnant 60 days after embryo transfer two aborted dead foetuses, at 4 and 9.5 months, and one gave birth to a live calf. Drost, Wright & Elsden (1986) investigated embryo transfer between water buffalo (*Bubalus bubalus*) and domestic cattle, but no pregnancies were obtained.

Embryo transfer has been used to propagate the bongo antelope (*Tragelaphus euryceros*) at Cincinnati Zoo. Superovulation and non-surgical

Table 1. Pregnancy rates after surgical transfer of Przewalski's horse and Grant's zebra embryos to domestic horse and donkey recipients.

Embryo genotype	No. embryos transferred	Recipients		Pregnancies established initially[a]		Pregnancies maintained to term	
		Horses	Donkeys	Horses	Donkeys	Horses	Donkeys
E. przewalskii	9	9	—	7(78%)	—	4(44%)[b]	—
E. burchelli	12	5	7	3(60%)	2(29%)	2(40%)[b]	—

[a] Recipients in which a conceptus could be detected by ultrasound examination of the uterus on Days 16–20 were classified as having a pregnancy established initially.
[b] Percentage of embryos transferred.

Data from Summers, Shephard, Hodges *et al.* (1987).

embryo collection techniques were developed first in the eland (*Tragelaphus oryx*) (Dresser, Kramer, Pope, Dahlhausen & Blauser 1982) and then applied to the bongo. Of four bongo embryos transferred to eland one gave rise to a live calf which was successfully reared by its surrogate mother (Dresser, Pope, Kramer, Kuehn, Dahlhausen, Maruska, Reece & Thomas 1985).

Techniques for superovulation, induction of oestrus, and non-surgical embryo collection were described for several *Oryx* spp. (Durrant 1983; Wildt, Schiewe, Schmidt, Goodrowe, Howard, Phillips, O'Brien & Bush 1986), but there have been no reports of resulting pregnancies. Embryo collection and intra-specific transfer has also been successful in the llama (*Llama glama*) (Wiepz & Chapman 1985).

Live foals of the common zebra (*Equus burchelli*) (Bennett & Foster 1985) and of Przewalski's horse (*Equus przewalskii*) have been produced (Table 1) by transferring embryos to domestic horse recipients but, contrary to expectations, the domestic donkey was found unsuitable as a recipient for zebra embryos by Summers, Shephard, Hodges, Kydd, Boyle & Allen (1987). These authors showed that six to eight embryos could be collected from a donor mare during a breeding season and that two or three foals could be obtained each year. The complete genetic disparity between the conceptus and the surrogate mother in these equid pregnancies raises the question as to whether immunological rejection by a surrogate mother is the cause of pregnancy failures. A heightened maternal immune response to foetal antigens was not related to success or failure of pregnancy (Summers, Shephard, Hodges *et al.* 1987).

Embryo transfer in non-human primates

Offspring have been born from a number of species of non-human primates as a result of embryo transfer, but the recipients were usually of the same species as the donors (Kraemer, Moore & Kramen 1976; Marston, Penn &

Sivelle 1977; Pope, Pope & Beck 1983; Hodgen 1983; Balmaceda, Pool, Arana, Heitman & Asch 1984; Bavister, Boatman, Collins, Dierschke & Eisele 1984; Summers, Shephard, Taylor & Hearn 1987). These studies, which were confined to laboratory primates, formed part of biomedical research programmes directed towards a better understanding of the normal embryological, endocrinological and immunological events that occur during early pregnancy in primates including man.

Regulation of the reproductive capacity of exotic species by embryo transfer technology has so far been applied only to animals in zoos. Besides having the potential to increase the number of animals of a particular species, the techniques can also be used selectively within defined genetic management programmes to increase the fecundity of selected animals. They also open the way for the transfer of genetic material in the form of embryos rather than of live animals within and between countries, a trade already in operation for frozen cattle embryos (Voss, Landmann, Wilke, Hasler, Whitaker & Heilman 1986). It may be possible to collect embryos from animals such as the bongo antelope in the wild habitat (Dresser 1986) to strengthen the gene pool of captive populations, but animal health authorities must be satisfied that exotic diseases will not be introduced with the embryos.

Embryo transfer and disease control

The widespread application of embryo transfer in domestic animal breeding is accompanied by some concern that diseases, principally those of viral aetiology, may be spread by the technique. On the other hand, the optimists propose that embryo transfer may be a means of controlling certain diseases, but until recently there has been only limited knowledge of viral infections of pre-implantation mammalian embryos on which to base a sound scientific view.

Infection of an embryo might arise via the maternal or paternal genome; or by contamination of the oocyte or spermatozoon; or from an infectious agent within the oviduct or uterus. There is evidence that the ovary can become infected with viruses. In cattle, bovine viral diarrhoea-mucosal disease virus has been shown to cause diffuse interstitial ovaritis (Ssentongo, Johnson & Smith 1980) and akabane virus injected into the uterus of heifers at oestrus causes local necrosis and inflammation of the corpus luteum (Parsonson, Della-Porta & Snowdon 1981), but it is not known whether oocytes become infected. There is probably a greater likelihood of the spread of infection—at least in cattle—from semen than from an infected ovary. In cattle many viruses including those of bluetongue, infectious bovine rhinotracheitis and foot-and-mouth disease can be present in semen (Kahrs, Gibbs & Larson 1980).

The intact zona pellucida—the glycoprotein coat surrounding the embryo—has been shown experimentally to be an effective barrier to many viruses and bacteria, but the physical size of an infectious agent may be important. Thus, in the mouse, the mengovirus, a small RNA virus, can pass down the corona cell tracts (Gwatkin 1967). However, viruses may attach to the zona pellucida (Singh, Thomas, Papp-Vid, Eaglesome & Hare 1982) and sometimes can be found deep in the spermatozoon tracts of the zona (Bolin, Turek, Runnels & Gustafson 1983). Bluetongue virus, an arthropod-borne orbivirus that infects domestic and wild ruminants, is highly embryocidal to zona-free cattle embryos (Bowen, Howard & Pickett 1982). Despite this, transfer of embryos collected from cows at the peak of bluetongue viraemia did not infect recipients (Bowen, Howard, Elsden & Seidel 1983). Similarly, recipients receiving embryos from donors infected with bovine leukaemia virus did not develop leukaemia antibodies nor did their calves (Hare, Mitchell, Singh, Bouillant, Eaglesome, Ruckerbauer, Bielanski & Randall 1985). Embryo transfer is capable of conserving valuable genetic material from pig herds infected with pseudorabies virus (James, James, Martin, Reed & David 1983). However, pig embryos exposed to pseudorabies virus *in vitro* or from experimentally infected sows transmitted the virus to recipients but not to their offspring (Bolin, Runnels, Sawyer & Gustafson 1982). In laboratory mice, Sendai (parainfluenza type 1) virus was shown not to be transmitted from infected embryo donors to recipients or their progeny (Carthew, Wood & Kirby 1983).

Current scientific opinion suggests that, while it is difficult to rely on the inherent sterility of the embryo, disease transmission is less likely to arise from embryo transfer carried out with appropriate precautions (Singh 1985) than from the movement of live animals. Nevertheless, the animal health authorities of a country importing embryos would still require rigorous tests for disease in the embryo donors, and both testing and quarantine of the recipients and progeny. It would seem unlikely that, in the foreseeable future, embryos from exotic species, particularly those of wild ungulates, will be allowed into those countries free of the major diseases of livestock without stringent and costly quarantine procedures.

Embryo cryopreservation

After the successful development of embryo freezing in mice (Whittingham, Leibo & Mazur 1972; Wilmut 1972), numerous reports have described embryo freezing and storage in liquid nitrogen in other species, particularly in domestic ruminants (see Leibo 1981). Embryo freezing, which is an integral part of the cattle embryo transfer industry, is used to store excess embryos when insufficient recipients are available. An increasingly important application is the national and international transport of cattle embryos frozen in liquid nitrogen, a procedure that is easier, cheaper and more humane than the transport of the live animal.

Embryos have been frozen from several exotic ungulates including the gaur cow (Stover & Evans 1984), Arabian oryx (Wildt *et al.* 1986), eland (Dresser, Kramer, Dahlhausen, Pope & Baker 1984) and bongo (Dresser, Pope *et al.* 1985) but so far the only live birth has been that of an eland calf (Dresser, Kramer, Dahlhausen *et al.* 1984). Embryos have also been successfully frozen from several non-human primates, including the baboon (Pope, Pope & Beck 1984), cynomolgus monkey (Balmaceda, Heitman, Garcia, Pauerstein & Pool 1986), and the common marmoset (Summers, Shephard, Taylor *et al.* 1987).

There are several advantages in being able to store embryos from exotic species in liquid nitrogen. One appealing concept is the development of embryo banks as a precaution against the loss of a species through accident or disease. Ultimately, it may be possible to freeze oocytes; these could later be thawed, matured and fertilized *in vitro*, and the embryo transferred to a suitable recipient. As in cattle, the freezing of embryos of exotic species will be useful for logistical reasons in embryo transfer programmes. It may also be possible, assuming disease control measures are observed, to transport embryos between countries.

Embryo manipulation for the study of mechanisms of disease resistance

It has long been recognized that there are obvious differences between breeds of cattle in their susceptibility and resistance to specific diseases and parasitic infestations, particularly those endemic in tropical areas of the world. In Northern Australia, *Bos indicus* breeds of cattle are more resistant than *B. taurus* breeds to infectious bovine keratoconjunctivitis (Frisch 1975) and to infestation with the calf tick, *Boophilus microplus* (Utech, Wharton & Kerr 1978); and the rustic N'Dama and Maturu cattle of West Africa are more resistant than introduced *Bos taurus* and *B. indicus* to trypanosomiasis (Desomitz 1959; Murray, Morrison, Murray, Clifford & Trail 1979). It is not known whether such differences are dependent on the immune system or on other physiological or anatomical features of the animal. For instance, is tick resistance related to the special features of *Bos indicus* skin and subcutaneous tissue or to a particularly efficient immune rejection of the attaching ticks?

It has been proposed that the role of the genetic constitution of the immune system in determining the differences in disease resistance between *Bos indicus* and *B. taurus* cattle could be examined experimentally in *Bos indicus–Bos taurus* haemopoietic (and thus lymphoid) chimeric twin calves (Summers, Shelton, Morris & Bell 1984). Such calves (Fig. 1) were

Fig. 1. A pair of Brahman–Jersey haemopoietic chimeric calves (GC16, GC15) with their Jersey mother.

produced by transferring Day 6–8 embryos from *Bos taurus* (Jersey, Friesian and Angus) and *B. indicus* (Brahman) donors to the contralateral uterine horn of previously inseminated *B. indicus* and *B. taurus* recipients respectively (Summers, Shelton & Edwards 1983). In cattle, though not in most other species, twin calves are haemopoietic chimeras because the chorioallantoic blood vessels of the conceptuses join during the second month of gestation (Hafez & Rajakoski 1964) permitting the interchange of haemopoietic stem cells. All the Brahman–Friesian and Angus–Brahman twin calves produced by embryo transfer in these experiments were haemopoietic chimeras but interestingly only 50% of the Brahman–Jersey twins were chimeras. The unique quality of these animals was that most had a lymphocyte population that was predominantly of one genotype (Table 2). The responsiveness of the immune apparatus to defined antigens and disease agents can be studied in such animals by the chronic catheterization of the efferent lymphatic ducts of a peripheral lymph node such as the prefemoral lymph node (Fig. 2) (Hall & Morris 1962; Summers, Shelton, Morris *et al.* 1984).

The concept of using cattle chimeras for the study of disease resistance mechanisms could be extended to whole-body chimeras derived by the reconstruction of pre-implantation embryos with embryonic cells of the two breeds concerned (Summers, Shelton & Bell 1983; Brem, Tenhumberg & Kräusslich 1984; Munro, Williams & Shelton 1986). With this model,

Table 2. The proportion of *Bos indicus* and *Bos taurus* lymphocytes in the pre-femoral efferent lymph of *Bos indicus–Bos taurus* twin calves.

Twin no.	Animal no.	Percentage *Bos indicus* lymphocytes	Percentage *Bos taurus* lymphocytes
1	Friesian SV4[a]	4.9	95.1
2	Angus SV7[a]	9.9	90.1
3	Friesian SV8	0.6	99.4
	Brahman SV9	2.3	97.7
4	Friesian SV15	0.6	99.4
	Brahman SV16	0	100
5	Friesian SV17	4.9	95.1
	Brahman SV21	2.1	97.9
6	Friesian SV20	0	100
	Brahman SV21	0.9	99.1
7	Friesian SV24	1.7	98.3
	Brahman SV25	0.8	99.2
8	Jersey GC13	32.1	67.9
	Brahman GC14	55.4	44.6
9	Jersey GC15	99.2	0.8
	Brahman GC16	100	0

[a] Brahman co-twin died before testing.

Data from Summers, Shelton, Morris *et al.* (1984).

much could be learnt about the mechanism of cattle tick resistance by sequential infestations of *Bos indicus* and *B. taurus* skin in a chimeric animal.

Embryo transfer technology has provided a means to study the basic mechanisms of trypanotolerance in disease-free N'Dama cattle at the International Laboratory for Research on Animal Diseases in Kenya (Jordt, Mahon, Tournay, Ngulo, Morrison, Rawle & Murray 1986). Embryos were collected from N'Dama in the Gambia, frozen and transported in liquid nitrogen to Kenya where they were transferred to Boran recipients. The successful birth of N'Dama calves once again highlighted the value of embryo manipulative techniques for the transport of genetic material over long distances and for the establishment of a herd free of a specific disease.

Embryo micromanipulation

The techniques of microsurgical manipulation of pre-implantation embryos have undergone major advances in recent years since their development for studies on experimental embryology in rodents (Gardner 1968; Lin 1969; Gardner 1971). Microsurgical division of morulae or early blastocysts is frequently used in cattle embryo transfer programmes (Ozil, Heyman &

Fig. 2. A Friesian calf in which the efferent lymphatic duct of the right prefemoral lymph node has been cannulated. The cannula is passed through the skin and the lymph collected into a detachable plastic tissue-culture flask.

Renard 1982; Lambert, Looney, Voelkel & Jackson 1983; Williams, Elsden & Seidel 1984; Takeda, Hallowell, McGauley & Hasler 1986). Depending on the techniques employed, up to 1.5 pregnancies per original intact embryo have been reported. Some of the calves produced by these methods are monozygotic twins and these animals are useful for certain research purposes. Division of embryos of mice, sheep, pigs and horses has also been reported but embryos up to the eight-cell stage have frequently been used (Willadsen 1980; Allen & Pashen 1984).

As mentioned above, embryo transfer in exotic species may necessitate the use of a female of another species as a surrogate mother. However, some species combinations are unsuccessful (Kraemer 1983). Failures are probably the result of immunological or endocrinological incompatibilities, or both, between the conceptus and the surrogate mother. The presence of the trophoblast cell layer at the interface of the conceptus and maternal tissues is believed to play a major role in the successful establishment and maintenance of pregnancy. It has been shown with *Mus musculus* and *M. caroli* and with sheep and goats, two species combinations in which reciprocal embryo transfer is unsuccessful, that the reproductive failure can be overcome by reconstructing embryos so that (1) all or some of the trophoblast *is* of the

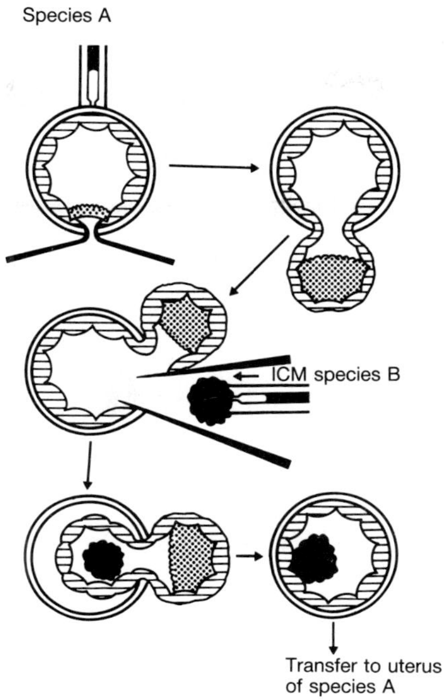

Fig. 3. Diagrammatic representation of the techniques used to replace the inner cell mass (ICM) of a blastocyst with that from a blastocyst of another species. The zona pellucida of the blastocyst of species A is opened with micro-instruments to allow the extrusion of the ICM. An isolated ICM from species B is placed into the species A blastocyst and its extruded ICM removed. The reconstructed blastocyst is transferred to the uterus of species A. (From Summers 1986.)

genotype of the surrogate mother, and (2) the inner cell mass—which forms the embryo proper—is of the genotype to be propagated (Rossant, Croy, Clark & Chapman 1983; Fehilly, Willadsen & Tucker 1984; Meinecke-Tillmann & Meinecke 1984). If such reconstruction techniques are to be used to aid the propagation of exotic species, it will be necessary to avoid the production of whole-body chimeras. The procedure represented diagrammatically in Fig. 3 is probably the best for avoiding the production of chimeras, although technically it is complex.

The desire to pre-determine the sex of offspring of man and animals is virtually as old as mankind itself. Biopsy techniques for obtaining embryo cells for the purpose of sex determination have been developed for several species (Fig. 4), but the success rate has generally been too low for their widespread use (Picard, King & Betteridge 1985). Non-invasive immunological methods for the detection of male specific antigens (White, Bradbury, Anderson & BonDurrant 1984), or metabolic analysis for enzymes coded

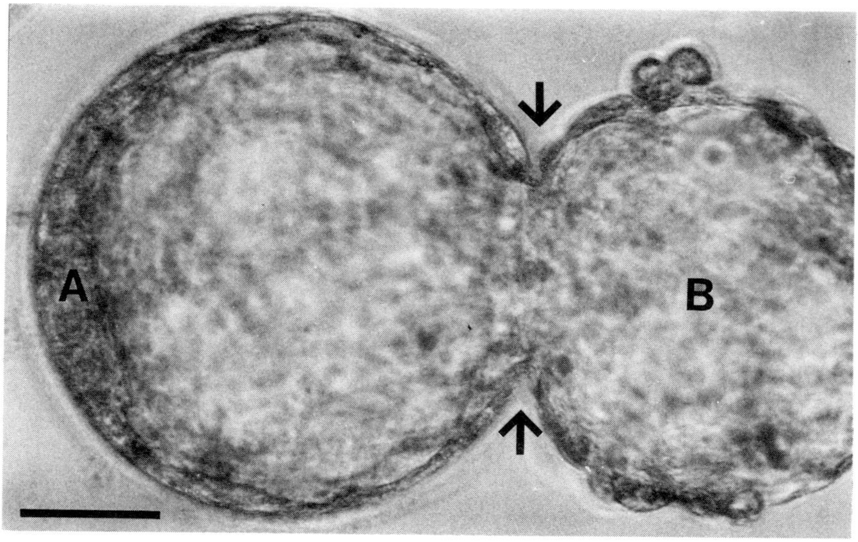

Fig. 4. A marmoset monkey blastocyst to illustrate that after opening the zona pellucida opposite the inner cell mass (A), trophectoderm will herniate (B) through the break in the zona after a period in culture. The herniated portion can be cut off (arrows) and the cells used to determine the sex of the embryo. Scale bar =50 μm.

on the X chromosome (Rieger 1984), may in the future prove to be more satisfactory than the biopsy of embryonic cells.

Acknowledgements

The work reported in this paper was supported by grants from the Australian Meat Research Committee, the Medical Research Council (UK) and the Agricultural and Food Research Council (UK).

References

Allen, W.R. (1982). Embryo transfer in the horse. In *Mammalian egg transfer*: 135–54. (Ed. Adams, C.E.). CRC Press, Florida.

Allen, W.R. & Pashen, R.L. (1984). Production of monozygotic (identical) horse twins by embryo micromanipulation. *J. Reprod. Fert.* **71**: 607–13.

Balmaceda, J.P., Heitman, T.O., Garcia, M.R., Pauerstein, C.J. & Pool, T.B. (1986). Embryo cryopreservation in cynomolgus monkeys. *Fert. Steril.* **45**: 403–6.

Balmaceda, J.P., Pool, T.B., Arana, J.B., Heitman, T.S. & Asch, R.H. (1984). Successful *in vitro* fertilization and embryo transfer in cynomolgus monkeys. *Fert. Steril.* **42**: 791–5.

Bavister, B.D., Boatman, D.E., Collins, K., Dierschke, D.J. & Eisele, S.G. (1984). Birth of rhesus monkey infant after *in vitro* fertilization and non-surgical embryo transfer. *Proc. natn. Acad. Sci. U.S.A.* **81:** 2218–22.

Bennett, S.D. & Foster, W.R. (1985). Successful transfer of a zebra embryo to a domestic horse. *Equine vet. J. Suppl.* **3:** 78–9.

Bolin, S.R., Runnels, L.J., Sawyer, C.A. & Gustafson, D.P. (1982). Experimental transmission of pseudorabies virus in swine by embryo transfer. *Am. J. vet. Res.* **43:** 278–80.

Bolin, S.R., Turek, J.J., Runnels, L.J. & Gustafson, D.P. (1983). Pseudorabies virus, porcine parvovirus and porcine enterovirus interactions with the zona pellucida of the porcine embryo. *Am. J. vet. Res.* **44:** 1036–9.

Bowen, R.A., Howard, T.H., Elsden, R.P. & Seidel, G.E. (1983). Embryo transfer from cattle infected with bluetongue virus. *Am. J. vet. Res.* **44:** 1625–8.

Bowen, R.A., Howard, T.H. & Pickett, B.W. (1982). Interaction of bluetongue virus with preimplantation embryos from mice and cattle. *Am. J. vet. Res.* **43:** 1907–11.

Brem, G., Tenhumberg, H. & Kräusslich, H. (1984). Chimerism in cattle through microsurgical aggregation of morulae. *Theriogenology* **22:** 609–13.

Bunch, T.D., Foote, W.C. & Whitaker, B. (1977). Interspecies ovum transfer to propagate wild sheep. *J. Wildl. Mgmt* **41:** 726–30.

Carthew, P., Wood, M.J. & Kirby, C. (1983). Elimination of Sendai (parainfluenza type 1) virus infection from mice by embryo transfer. *J. Reprod. Fert.* **69:** 253–7.

Desomitz, R.S. (1959). Studies on immunity and host-parasite relationships. 1. The immunological responses of resistant and susceptible breeds of cattle to trypanosomal challenge. *Ann. trop. Med. Parasit.* **53:** 293–313.

Dresser, B.L. (1986). Embryo transfer in exotic bovids. *Int. Zoo Yb.* **24/25:** 138–42.

Dresser, B.L., Kramer, L., Dahlhausen, R.D., Pope, C.E. & Baker, R.D. (1984). Cryopreservation followed by successful transfer of African eland antelope (*Tragelaphus oryx*) embryos. *Proc. int. Congr. Anim. Reprod. artif. Insem.* **10**(2): 191. (Abstr.)

Dresser, B.L., Kramer, L., Pope, C.E., Dahlhausen, R.D. & Blauser, C. (1982). Superovulation of African eland (*Taurotragus oryx*) and interspecies embryo transfer to Holstein cattle. *Theriogenology* **17:** 86. (Abstr.)

Dresser, B.L., Pope, C.E., Kramer, L., Kuehn, G., Dalhausen, R.D., Maruska, E.J., Reece, B. & Thomas, W.D. (1985). Birth of bongo antelope (*Tragelaphus euryceros*) to eland antelope (*Tragelaphus oryx*) and cryopreservation of bongo embryos. *Theriogenology* **23:** 190. (Abstr.)

Drost, M., Wright, J.M. & Elsden, R.P. (1986). Intergeneric embryo transfer between water buffalo and domestic cattle. *Theriogenology* **25:** 13–23.

Durrant, B.S. (1983). Reproductive studies of the oryx. *Zoo Biol.* **2:** 191–7.

Durrant, B.S. & Benirschke, K. (1981). Embryo transfer in exotic animals. *Theriogenology* **15:** 77–83.

Edwards, R.G. (1985). Current status of human conception *in vitro*. *Proc. R. Soc.* (B)**223:** 417–48.

Fehilly, C.B., Willadsen, S.M. & Tucker, E.M. (1984). Interspecific chimaerism between sheep and goat. *Nature, Lond.* **307:** 634–6.

Frisch, J.W. (1975). The relative incidence and effect of bovine keratoconjunctivitis in *Bos indicus* and *Bos taurus* cattle. *Anim. Prod.* **21:** 265–74.

Gardner, R.L. (1968). Mouse chimaeras obtained by the injection of cells into the blastocyst. *Nature, Lond.* **220:** 596–7.

Gardner, R.L. (1971). Manipulations on the blastocyst. *Adv. Biosci.* **6:** 279–96.

Goodrowe, K.L., Howard, J.G. & Wildt, D.E. (1986). Embryo recovery and quality in the domestic cat: natural versus induced ovulation. *Theriogenology* **25:** 156. (Abstr.)

Gwatkin, R.B.L. (1967). Passage of mengovirus through the zona pellucida of the mouse morula. *J. Reprod. Fert.* **13:** 577–8.

Hafez, E.S.E. & Rajakoski, E. (1964). Placental and fetal development during multiple bovine pregnancy. Anatomical and physiological studies. *Anat. Rec.* **150:** 303–16.

Hall, J.G. & Morris, B. (1962). The output of cells in the lymph from the popliteal node of sheep. *Q. Jl. exp. Physiol.* **47:** 360–9.

Hare, W.C.D., Mitchell, D., Singh, E.L., Bouillant, A.M.P., Eaglesome, M.D., Ruckerbauer, G.M., Bielanski, A. & Randall, G.C.B. (1985). Embryo transfer in relation to bovine leukemia virus control and eradication. *Can. vet. J.* **26:** 231–4.

Hodgen, G.D. (1983). Surrogate embryo transfer combined with estrogen-progesterone therapy in monkeys. *J. Am. med. Ass.* **250:** 2167–71.

James, J.E., James, D.M., Martin, P.A., Reed, D.E. & David, D.L. (1983). Embryo transfer for conserving valuable genetic material from swine herds with pseudorabies. *J. Am. vet. med. Ass.* **183:** 525–8.

Jordt. T., Mahon, G.D., Tournay, B.N., Ngulo, W.K., Morrison, W.I., Rawle, J. & Murray, M. (1986). Successful transfer of N'Dama embryos into Boran recipients. *Vet. Rec.* **119:** 246–7.

Kahrs, R.F., Gibbs, E.P.J. & Larsen, R.E. (1980). The search for viruses in bovine semen, a review. *Theriogenology* **14:** 151–65.

Kraemer, D.C. (1983). Intra- and interspecific embryo transfer. *J. exp. Zool.* **228:** 363–71.

Kraemer, D.C., Moore, G.T. & Kramen, M.A. (1976). Baboon infant produced by embryo transfer. *Science, N.Y.* **192:** 1246–7.

Lambert, V.A., Looney, C.R., Voelkel, S.A. & Jackson, D.A. (1983). Microsurgery on bovine embryos at the morula stage to produce monozygotic twin calves. *Theriogenology* **20:** 85–95.

Leibo, S.P. (1981). Preservation of ova and embryos by freezing. In *New technologies in animal breeding:* 127–39. (Eds Brackett, B.G., Seidel, G.E. & Seidel, S.M.). Academic Press, New York.

Lin, T.P. (1969). Microsurgery of inner cell mass of mouse blastocysts. *Nature, Lond.* **222:** 480–1.

Marston, J.H., Penn, R. & Sivelle, P.C. (1977). Successful autotransfer of tubal eggs in the rhesus monkey, *Macaca mulatta. J. Reprod. Fert.* **49:** 175–6.

Meinecke-Tillmann, S. & Meinecke, B. (1984). Experimental chimaeras–removal of reproductive barrier between sheep and goat. *Nature, Lond.* **307:** 637–8.

Munro, R.K., Williams, T.J. & Shelton, J.N. (1986). Production of overt interspecies bovine chimaeras by aggregation of demi-embryos. *Proc. a. Conf. Aust. Soc. reprod. Biol.* **8:** 15. (Abstr.).

Murray, M., Morrison, W.I., Murray, P.K., Clifford, D.J. & Trail, J.C.M. (1979). Trypanotolerance—a review. *Wld Anim. Rev.* **31:** 2–12.

Ozil, J.-P., Heyman, Y. & Renard, J.-P. (1982). Production of monozygotic twins by micromanipulation and cervical transfer in the cow. *Vet. Rec.* **110:** 126–7.

Parsonson, I.M., Della-Porta, A.J. & Snowdon, W.A. (1981). The consequences of infection of cattle with Akabane virus at the time of insemination. *J. comp. Path. Ther.* **91:** 611–9.

Picard, L., King, W.A. & Betteridge, K.J. (1985). Production of frozen sexed bovine embryos by micromanipulation and cytogenetics. *Vet. Rec.* **117:** 603–8.

Pope, C.E., Pope, V.Z. & Beck, L.R. (1983). Successful nonsurgical transfer of a non-surgically recovered four-cell uterine embryo in the baboon. *Theriogenology* **19:** 144. (Abstr.)

Pope, C.E., Pope, V.Z. & Beck, L.R. (1984). Live birth following cryopreservation and transfer of a baboon embryo. *Fert. Steril.* **42:** 143–5.

Rieger, D. (1984). The measurement of metabolic activity as an approach to evaluating viability and diagnosing sex in early embryos. *Theriogenology* **21:** 138–40.

Rossant, J., Croy, B.A., Clark, D.A. & Chapman, V.M. (1983). Interspecific hybrids and chimeras in mice. *J. exp. Zool.* **228:** 223–33.

Seidel, G.E. & Seidel, S.M. (1981). The embryo transfer industry. In *New techniques in animal breeding*: 41–80. (Eds Brackett, B.G., Seidel, G.E. & Seidel, S.M.). Academic Press, New York.

Singh, E.L. (1985). Disease control: procedures for handling embryos. *Rev. scient. tech. Off. int. Epizoot.* **4:** 867–72.

Singh, E.L., Thomas, F.C., Papp-Vid, G., Eaglesome, M.D. & Hare, W.C.D. (1982). Embryo transfer as a means of controlling the transmission of viral infections. II. The *in vitro* exposure of preimplantation bovine embryos to infectious bovine rhinotracheitis virus. *Theriogenology* **18:** 133–40.

Ssentongo, Y.K., Johnson, R.H. & Smith, J.R. (1980). Association of bovine viral diarrhoea–mucosal disease virus with ovaritis in cattle. *Aust. vet. J.* **56:** 272–3.

Stover, J. & Evans, J. (1984). Interspecies embryo transfer from Gaur (*Bos gaurus*) to domestic Holstein cattle (*Bos taurus*) at the New York Zoological Park. *Proc. int. Congr. Anim. Reprod. artif. insem.* **10**(2): 243.(Abstr.)

Stover, J., Evans, J. & Dolensek, E.P. (1981). Interspecies embryo transfer from the Gaur to domestic Holstein. *A. Proc. Am. Ass. Zoo Vet.* **1982:** 122–4.

Stover, J. & Westrom, W.K. (1984). Reproductive Studies Unit of the New York Zoological Society. *Zoo Biol.* **3:** 335–41.

Summers, P.M. (1986). Collection, storage and use of mammalian embryos. *Int. Zoo. Yb.* **24/25:** 131–8.

Summers, P.M., Shelton, J.N. & Bell, K. (1983). Synthesis of primary *Bos taurus–Bos indicus* chimaeric calves. *Anim. Reprod. Sci.* **6:** 91–102.

Summers, P.M., Shelton, J.N. & Edwards, J. (1983). The production of mixed-species *Bos taurus–Bos indicus* twin calves. *Anim. Reprod. Sci.* **6:** 78–89.

Summers, P.M., Shelton, J.N., Morris, B. & Bell, K. (1984). Interspecific chimerism—the characterization and immunological responsiveness of *Bos taurus–Bos indicus* haemopoietic chimeras produced by embryo transfer. *Aust. J. exp. Biol. med. Sci.* **62:** 27–45.

Summers, P.M., Shephard, A.M., Hodges, J.K. Kydd, J., Boyle, M.S. & Allen, W.R. (1987). Successful transfer of the embryos of Przewalski's horse (*Equus przewalskii*) and Grant's zebra (*E. burchelli*) to domestic mares (*E. caballus*). *J. Reprod. Fert.* **80:** 13–20.

Summers, P.M., Shephard, A.M., Taylor, C.T. & Hearn, J.P. (1987). The effects of cryopreservation and transfer on embryonic development in the common marmoset monkey. *J. Reprod. Fert.* **79:** 241–50.

Takeda, T., Hallowell, S.V., McGauley, A.D. & Hasler, J.F. (1986). Pregnancy rates with intact and split bovine embryos transferred surgically and non-surgically. *Theriogenology* **25:** 204. (Abstr.)

Utech, K.B.W., Wharton, R.H. & Kerr, J.D. (1978). Resistance to *Boophilus microplus* (Canestrini) in different breeds of cattle. *Aust. J. agric. Res.* **29:** 885–95.

Voss, H.J., Landmann, D., Wilke, G., Hasler, J.F., Whitaker, R.O. & Heilman, R.D. (1986). Pregnancy rate after surgical transfer of imported frozen bovine embryos. *Theriogenology* **25:** 209. (Abstr.)

White, K.L., Bradbury, M.W., Anderson, G.B. & BonDurrant, R.H. (1984). Immunofluorescent detection of a male-specific factor on preimplantation bovine embryos. *Theriogenology* **21:** 275.(Abstr.)

Whittingham, D.G., Leibo, S.P. & Mazur, P. (1972). Survival of mouse embryos frozen to −196°C and −269°C. *Science, N.Y.* **178:** 411–4.

Wiepz, D.W. & Chapman, R.J. (1985). Non-surgical embryo transfer and live birth in a llama. *Theriogenology* **24:** 251–7.

Wildt, D.E., Schiewe, M.C., Schmidt, P.M., Goodrowe, K.L., Howard, J.G., Phillips, L.G., O'Brien, S.J. & Bush, M. (1986). Developing animal model systems for embryo technologies in rare and endangered wildlife. *Theriogenology* **25:** 33–51.

Willadsen, S.M. (1980). The viability of early cleavage stages containing half the normal number of blastomeres in the sheep. *J. Reprod. Fert.* **59:** 357–62.

Williams, T.J., Elsden, R.P. & Seidel, G.E. (1984). Pregnancy rates with bisected bovine embryos. *Theriogenology* **22:** 521–31.

Wilmut, I. (1972). The effect of cooling rate, warming rate, cryoprotective agent and stage of development on survival of mouse embryos during freezing and thawing. *Life Sci.* **11:** 1071–9.

Symp. zool. Soc. Lond. (1988) No. 60 135–150

Diagnosis of pregnancy and disease by ultrasound in exotic species

G. H. du BOULAY
and O. L. WILSON

Nuffield Laboratories of Comparative Medicine
Institute of Zoology
The Zoological Society of London
Regent's Park
London NW1 4RY, UK

Synopsis

The development of real-time ultrasound at the Zoological Society of London is described. It has been directed to the monitoring of ovarian function and pregnancy, the diagnosis of pregnancy and disease, and the sexing of monomorphic species.

Experience in the use of a variety of probes for different purposes in mammals and reptiles of a range of sizes is reported. Some of the findings are described.

The value of real-time ultrasound in zoos is discussed. It is concluded that these techniques have already proved uniquely helpful, especially in the breeding programmes.

Disease diagnosis with ultrasound in exotic species is in its infancy, but may prove to be at least as important as conventional radiology.

Introduction

Radiological diagnosis of disease by X-rays has been a part of the work of the Hospital of The Zoological Society of London for many years. In 1965 The Wellcome Trust equipped a new Radiology Department in the then Nuffield Institute of Comparative Medicine (now part of the Institute of Zoology) so that radiological knowledge derived from human medicine could be used for the benefit of animals, while at the same time examination of animals might throw light on human medical problems. Shortly thereafter The Wellcome Trust also founded the Wellcome Animal X-Ray Museum within the Nuffield Institute.

Collaboration between the radiological facility in the Nuffield Institute and that in the Zoo Hospital, begun while Mr Oliver Graham-Jones was the Veterinary Officer, continued with Mr Malcolm Hime, DVR, when the first steps were taken to try to interrelate the radiological records.

As with X-rays, so also with real-time ultrasound, another radiological technique with which pioneering studies of primates and a few other zoo

ZOOLOGICAL SYMPOSIUM No. 60
ISBN 0-19-854007-8

animals have been carried out by James *et al.* (1976), Michejda & Hodgen (1981) and Nyland *et al.* (1984). A good, functional, but no longer state-of-the-art apparatus was first given to the Zoo's hospital. The use of this was explored jointly, particularly by Mr John Knight, veterinary officer, Mr Tony Fitzgerald, chief animal nurse, and Mrs Victoria Aitken, radiographer from the Nuffield Laboratories. It was soon apparent that real-time ultrasound could play a part in the diagnosis of pregnancy in circumstances where clinical observations and hormone assays were uncertain guides. Ultrasound succeeded in showing beyond doubt that a female giant panda (*Ailuropoda melanoleuca*) was not pregnant.

One of the authors (G.H.duB.), with much help from Mrs Aitken, introduced real-time ultrasound with modern apparatus into a research project concerned with reproduction in the common marmoset (*Callithrix jacchus*). To gain experience by working with so small an animal may seem strange, but the choice was fortunate. Common marmosets required no sedation or shaving, yet demanded a high standard of spatial and contrast resolution.

From this beginning the use of real-time ultrasound in the Zoological Society's collections at Regent's Park and Whipsnade has extended to pregnancy monitoring in other species and from obstetrics to gynaecological disease and to organs other than those of reproduction. Information on the diagnosis of disease and pregnancy in most species is still sparse, but from the common marmoset quite substantial numerical data have now accrued.

A further specialized use of ultrasound in certain monomorphic species is as a non-invasive means of determining sex.

Material

Animals examined for obstetric reasons

The animals (Table 1) were examined in order to diagnose, exclude or monitor pregnancy or, in a few cases, to seek reasons for failure to breed.

Animals examined for disease

Investigations undertaken because of failure to breed often revealed gynaecological disease. A wider search occasionally revealed other abnormalities. Clinical evidence sometimes led to concentrated examination of a particular organ, but often a search of other accessible areas was made in addition.

Animals examined for disease, or to establish normal appearances, are presented in Table 2.

Methods

Real-time ultrasound equipment with 3.5 MHz or 5 MHz transducers (see Tables 1 and 2) was used. Apparatus employed was supplied by Siemens

Table 1. Animals examined for obstetric reasons.

Species	Numbers examined	Transducers used
MAMMALS		
Grey short-tailed opossum *(Monodelphis domestica)*	9	5 MHz linear 5 MHz small-parts
Gorilla *(Gorilla gorilla)*	1	3.5 MHz sector
Orang utan *(Pongo pygmaeus)*	2	3.5 MHz sector
Chimpanzee *(Pan troglodytes)*	3	3.5 MHz sector
Common marmoset *(Callithrix jacchus)*	*c*.50	5 MHz linear (others tried)
Dolphin *(Tursiops truncatus)*	2	3.5 MHz straight and curved linear 3.5 MHz sector
Aardvark *(Orycteropus afer)*	3	5 and 3.5 MHz sector
African elephant *(Loxodonta africana)*	1	3.5 MHz sector (intrarectal)
Giraffe *(Giraffa camelopardalis)*	1	3.5 MHz sector
Kudu *(Tragelaphus strepsiceros)*	1	3.5 MHz sector
Waterbuck *(Kobus ellipsiprymnus)*	2	3.5 MHz sector
Bongo *(Boocerus euryceros)*	1	3.5 MHz sector
Scimitar horned oryx *(Oryx tao)*	5	3.5 MHz sector 5 MHz small-parts
Blackbuck *(Antilopa cervicapra)*	6	5 MHz sector
Camel *(Camelus bactrianus)*	1	3.5 MHz sector
REPTILES		
Common iguana *(Iguana iguana)*	3	5 MHz linear
Thailand water dragon *(Physignathus cocincinus)*	1	5 MHz linear
Eastern water dragon *(Physignathus lesueurii)*	1	5 MHz linear
Children's python *(Liasis childreni)*	1	5 MHz linear

Species within Orders have been listed in order of size—an important consideration with ultrasound. Unless otherwise stated 'linear' probes were straight.

Table 2. Animals examined for disease diagnosis.

Species	Numbers examined	Transducers used
Gorilla *(Gorilla gorilla)*	1	3.5 MHz sector
Lion-tailed macaque *(Macaca silenus)*	4	5 MHz sector
Dolphin *(Tursiops truncatus)*	2	3.5 MHz straight and curved linear 3.5 MHz sector
Killer whale *(Orcinus orca)*	2	2.5 MHz and 3.5 MHz straight linear
Giant panda *(Ailuropoda melanoleuca)*	1	3.5 MHz sector
Gemsbok *(Oryx gazella)*	1	3.5 MHz sector 5 MHz small-parts
Carpet python *(Morelia spilotus variegata)*	1	5 MHz linear
Black mamba *(Dendroaspis polylepis)*	1	5 MHz linear
Grey rat snake *(Elaphe obsoleta spiloides)*	1	5 MHz linear

Ltd. (Sonoline SL, used most often, and a Siemens portable), by Picker International Ltd. (5000 and 7000) and, for examination of a killer whale at Nice, by CGR (Compagnie Générale Radiologique).

The transducers usually available were 3.5 MHz linear, 5 MHz linear, and 3.5 MHz sector. Transducers on short trial included 5 MHz sector, 3.5 MHz curved linear and 5 MHz small-parts/rectal. A 3.5 MHz was used for the 'larger' animals as the lower frequency penetrates further than the higher. The 5 MHz was chosen for finer resolution in smaller animals. The size distinction was in practice found to lie, for example, between oryx and blackbuck in ungulates and between orang-utan and macaque in primates. The choice of transducer was generally determined by the size of the animal to be examined, though the quality of the tissue, which affects penetration and detail, also had to be taken into account.

In the animal's best interest sedation time had to be limited. This meant that there was often only time to pay attention to the more important diagnostic features. Some of the examinations were therefore far briefer than would otherwise have been desirable.

Examinations were carried out wherever convenient, except that every apparatus used required a mains electricity connection. Shelter from adverse weather conditions was sometimes necessary. When working out of doors the monitor screen was shaded from direct sunlight in order to see the image clearly. A number of different strategies were tried; for instance the monitor might be shaded by an umbrella or a hood, or detached and placed in the shadow of other apparatus or of people. Outdoor examinations were avoided when the animals could be moved to a roofed building; when this was impossible, however, a very long cable was used. An apparatus adapted to be driven from a car battery could also be used.

Even indoors there may be substantial hazards in examining aquatic animals such as dolphins and killer whales. The dolphins at Windsor Safari Park were winched up to surface level in their special pen; but killer whales, one at Windsor Safari Park and one in Nice, had to be examined free in their pools with all that that implied for electrical safety, discomfort of the ultrasonographer and possible disturbance due to the ultrasonic stimulus of the animal's hearing. The 2.5 MHz probe was sealed in a resinous case and attached to its instrument (manufactured by the CGR company) by an extra long cable, but the ordinary hospital (Siemens SX) instrument did not have such a lengthy probe lead and consequently had to be held by an assistant on a ladder just above the pool surface.

Preliminary trials showed that the 5 MHz linear probe, with a home-made jelly stand-off or a 'waterpath stand-off' supplied by the manufacturer, was the best available for marmosets. The function of the stand-off was to bring the reproductive tract to an optimum focal distance from the face of the probe, thus ensuring good resolution of the fine detail. Mechanical sector scanners were invaluable in taking advantage of any narrow window of approach through an area of soft fur or shaved skin, or in looking round the edge of a superficial bone. The 3.5 MHz probes were essential for surface examination of larger animals. For the largest, however, a 2.5 MHz probe was required to see an organ at any depth. Only limited experience was gained with small-parts probes, but unquestionably they are necessary for *early* pregnancy diagnosis in ungulates. The details of individual examinations are given later. The elephant (see below) was a special case in that a 3.5 MHz sector probe, protected by a plastic bag, was introduced rectally, by hand.

Sedation was used sparingly and only when necessary for the safety of the animal, handler or operator. None of the marmosets, for example, was sedated. Some animals, such as the bongo, had already been anaesthetized or sedated for other veterinary purposes. The need for sedation was not only determined by species: for instance some chimpanzees were successfully examined in familiar surroundings merely by restraint, whereas more temperamental members of the same species required sedation. The aardvarks were unsedated in their pen. The giraffe, at Longleat, was merely closely confined. The dolphins were restrained by hand, the whales merely coaxed. Most of the reptiles examined were small; without being harmed they were held in the air by practised handlers while the ultrasound probe was passed from below over the ventral surfaces of their bodies. The examination of a snake to determine its sex was also carried out in this way with little difficulty. On the other hand, because its bite was so poisonous, the black mamba was anaesthetized in a box, intubated, then maintained under anaesthesia for the entire examination.

The examination of birds' eggs for viability was complicated by the need to maintain contact with a sharply curved surface while avoiding contamination of the shell by fluid or gel. The precise method and the results of a controlled study are to be the subject of a further paper.

The obstacle presented by fur to the passage of the ultrasonic beam varied with both the animal and the part examined. For instance, most ungulates had an area of fine soft hair close to the nipples in the hypogastrium, which provided an excellent window when well impregnated with gel. Higher up on the abdominal wall shaving was often required. James *et al.* (1976) advocated shaving the body hair of primates to obtain good pictures. Yeager *et al.* (1981) also sometimes shaved their lowland gorillas. We have not found it essential in the examinations we have done.

Obstetric and gynaecological observations by ultrasound through the lower abdominal wall in human patients are usually carried out after a considerable water-load and without permitting the patient to micturate. The full bladder permits a very clear passage for the ultrasound beam. The method could not, however, be employed for animals, even primates, in which the examination techniques were otherwise similar to those used for women.

Findings

Diagnosis of pregnancy

The success of pregnancy diagnosis or exclusion by real-time ultrasound could, in our hands, be reliably assessed only in the common marmoset because we examined no other species in sufficient numbers. In marmosets a gestation sac could almost always be seen and measured by Day 16. Even before it was clearly visible changes occurred in the endometrial echo, which became a thick double line, strongly suggesting the likelihood of implantation, by Day 12.

In the opossums, although only a few were examined, embryos were visible even earlier (Day 3).

Very limited experience suggested that the features of early pregnancy in higher apes were like those in the human subject. Pregnancy is probably diagnosable with little doubt by 6 weeks. The experience of Yeager and his colleagues is greater than our own.

Only one dolphin was examined early in pregnancy: a foetus about

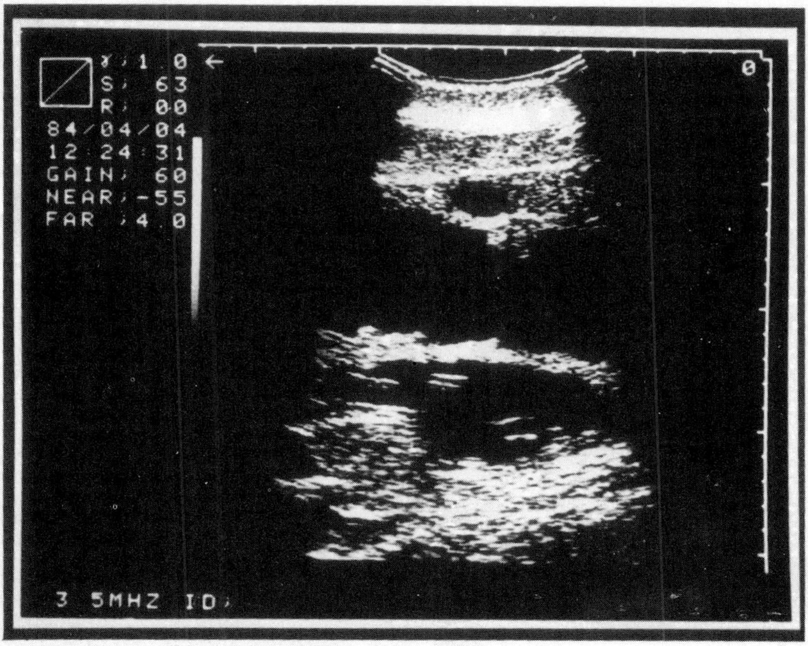

Fig. 1. Dolphin: early pregnancy. This image was obtained with a 3.5 MHz curved linear array probe, which explains why the ultrasound rays diverge. The white and grey areas at the top and bottom of the picture represent the mother's skin and blubber. Between these the black area is the amniotic fluid. In the centre of this, the white serrated shadow represents the posterior half of the spine of the foetus. The tail fluke is just out of sight on the right of the picture.

30 cm in length was unmistakably visualized (Fig. 1). The gestation period is 360–390 days and, by counting backwards from the date of birth, a foetal age of 130–160 days was calculated.

The pregnancy of one aardvark was followed to full term. At the first examination the foetal thorax had a cross-sectional diameter of 5 cm and an easily countable heart beat; retrospective assessment indicated its age to have been about 5 months.

From occasional observations of pregnancy in ungulates it was possible to show that pregnancy could be diagnosed by external real-time ultrasound examination in the giraffe (at 6 weeks), camel (at 12–14 weeks), bongo (at 180 days), scimitar-horned oryx (in the first third of pregnancy), and blackbuck (at 35 days). In the blackbuck, however, there was also one proved failure to diagnose a 35-day pregnancy. Our very limited experience with intrarectal scanning suggested that pregnancy could be diagnosed much earlier with a well designed small-parts probe.

Table 3. Relationship between foetal age and skull biparietal diameter (bpd) in the common marmoset.

Foetal age (days)	*N*	Mean bpd mm	S.D.
84–91	1	10.0	
92–98	5	11.0	1.58
99–105	6	14.6	1.21
106–112	2	16.5	
113–119	11	17.45	0.82
120–126	3	18.6	0.5
127–133	4	19.0	—
134–140	7	19.0	1.0
141–147	4	20.25	0.96

Monitoring of pregnancy

Monitoring of established pregnancy proved valuable in several ways. It may, as happened with the bongo, be reassuring to know that foetal development is continuing normally in spite of maternal illness requiring recurrent anaesthesia. Moreover, management may be aided by knowledge of the age and maturity of the foetus, when this can be established by reference to a data bank of ultrasound measurements derived from previous ultrasound examinations made with similar probes and validated against actual birth dates or knowledge of date of conception. It should be noted that linear and sector probes can apparently give different measurements (Nyland *et al.* 1984).

The common marmosets formed a group on their own as they were the subject of repeated examinations to monitor pregnancy under a variety of circumstances in the course of several different research projects. These animals are to be the subject of another communication. The accumulated data have been used to demonstrate, *inter alia*, the relationship between gestational age and foetal skull biparietal diameter. Table 3 is a record of the measurements from the pregnancies for which the day of ovulation was known with greatest precision from hormone assay.

The biparietal diameters of marmoset foetuses shown in Table 3 were used to confirm the clinical suspicion of post-maturity on at least three occasions. Evidence is now also beginning to suggest that in marmosets an unexpectedly small gestation sac or uterine cross-sectional area predicts impending foetal death. An observation made during the course of studies on chimpanzees suggested that in higher primates evidence of maturity during the last few weeks of pregnancy could be obtained from the echogenicity of the placenta, as happened with one of the chimpanzees (Fig. 2).

A further advantage of ultrasonic monitoring of pregnancy is that it may show a developmental abnormality, such as hydrocephalus (in a marmoset,

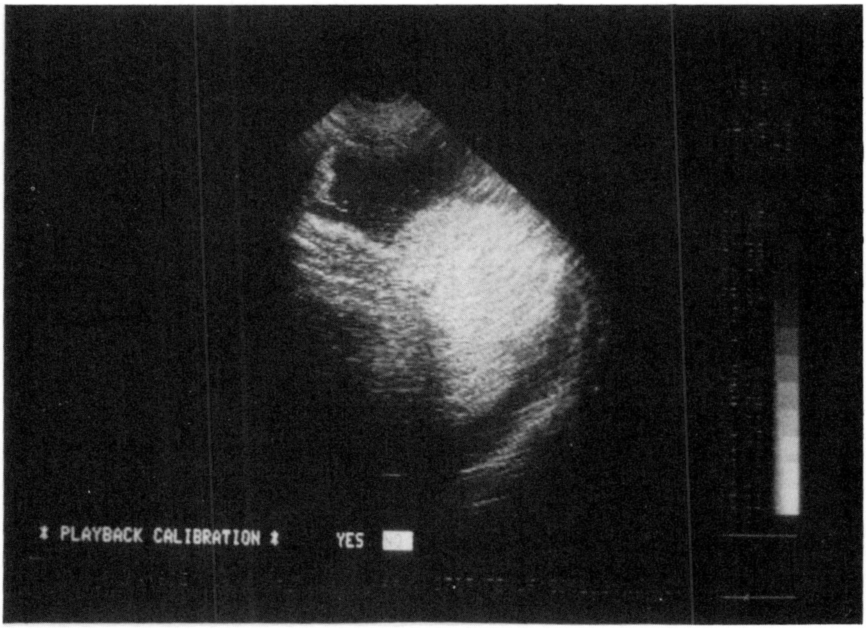

Fig. 2. Chimpanzee; mature placenta. A mechanical 3.5 MHz sector probe was used for this image. The area of contact on the skin is quite small and the divergence of the beam pronounced. The small curved segmental grey area at the top of the picture is caused by skin and subcutaneous tissues. Immediately deep to it is a black area of amniotic fluid. Impinging on this on the right is a large, bright white region representing the senescent, degenerating placenta.

in our experience), or the presence of multiple embryos, which has significant implications in some species.

Ovarian function

Many opportunities were taken to observe the ovaries of animals being examined both for possible pregnancy and for disease, but only in marmosets did we obtain sufficient experience to assess the uses and limitations of ultrasound in predicting impending ovulation, or revealing the status of the oestrous cycle. Marmoset follicles were so small that uncertainty remained up to about 25 h before ovulation, when two or three 3-mm follicles were often observed. During the middle part of the cycle, what appeared to be follicles of 1–2 mm diameter waxed and waned. Once a follicle reached 3 mm it could sometimes be watched by serial examination and its precise position on the surface of the ovary repeatedly checked.

In other mammals our first concern was to determine the position of the ovaries and how best to visualize them. In the giant panda, for instance, the diverging beam of the sector scanner was helpful in looking behind the bony brim in the lateral part of the pelvis.

Viability of eggs

Viability of eggs, in both birds and reptiles, was often difficult to establish very early, or if the shell was thick, or while the egg was still within the oviduct. We showed that ultrasound had some value. Snakes' eggs contained recognizable snakes. In a black mamba a swelling was shown not only to be an egg, but also to contain heterogeneous material. This non-viable egg was therefore surgically collapsed and removed. The essential clue to viability in a bird's egg was the beating heart. Ultrasound cannot yet be recommended, however, as a routine procedure for assessing the viability of eggs.

Sexing

The sexing of monomorphic snakes was successfully carried out by distinguishing the male cloacal pocket containing the hemipenis from that of the female. In the limited number of snakes examined no errors were made, in spite of the presence of a seminal plug.

Sexing of monomorphic mammals also seems possible from our experience with beavers. In the males the *os penis* was easily found.

Diagnosis of disease

Because of the way that the use of ultrasound has developed at the Zoo our experience of disease diagnosis is less than that of normal breeding.

Gynaecological disease

Endometriosis, in which numerous 'cysts' containing amorphous-looking debris were present in the enlarged uterus, was seen in one lion-tailed macaque (Fig. 3). A pyometrium was confirmed in another, in which the uterine cavity was enlarged and fairly uniformly trans-sonic, giving a darkish image compared with the uterine wall. Pyometrium was also seen in the marmoset.

A gemsbok with a rectally palpable mass was found to have an ovarian cyst with uniformly dark contents. Treated by prostaglandins the cyst changed, filling up with echogenic material. In the following year the animal came into season.

The liver

Apart from their size and position, normal mammalian livers look very similar (Fig. 4), and if the animal has a gall-bladder it is easily found. Because of their frequency we looked for, but did not find, liver abscesses in dolphins. In one sick killer whale, however, fatty degeneration was observed (Fig. 5).

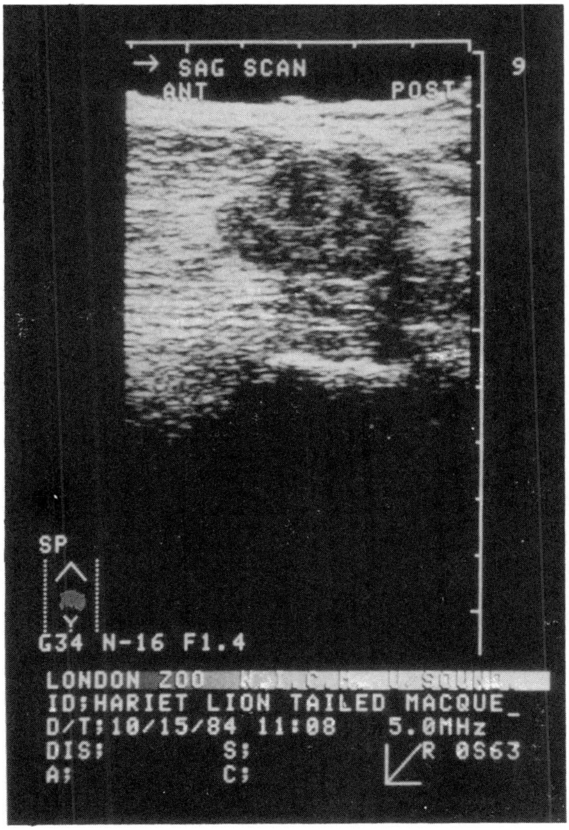

Fig. 3. Macaque: endometriosis. A 5 MHz straight linear array and a water-path standoff produced this image; 5 MHz ultrasound gives a more detailed picture for smaller animals. The centre of the image is occupied by a rounded non-homogeneous structure. This is the distended uterus, the cavity and walls of which are extensively involved with the changes of endometriosis. The appearance is typical of the advanced stage of the condition.

The gut

A female giant panda's chronic sterile peritonitis and ileus were clearly visualized as dilated and immobile loops of bowel with thick walls, separated by peritoneal fluid (Fig. 6).

The heart

The heart was routinely examined whenever possible, but only by real-time methods, rather than detailed movement analysis by M-mode. (Visualization of heart beat is, of course, an important part of the monitoring of pregnancy.)

An abnormal heart was observed in a snake with a localized swelling of

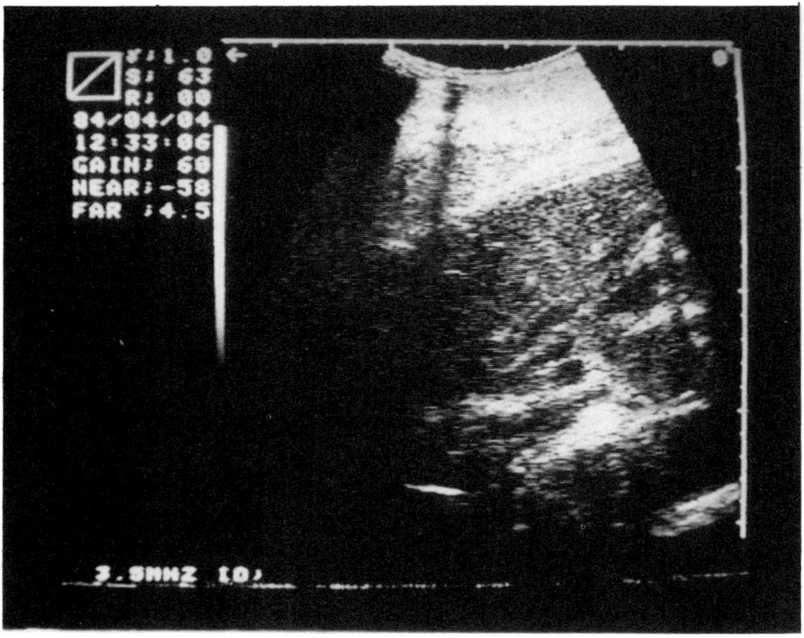

Fig. 4. Dolphin liver: normal. A 3.5 MHz curved linear array was used. The bright area at the top represents skin and blubber. The area deep to this is normal liver, showing arteries and veins. The round black hole surrounded by a wide white halo is probably a branch of the portal vein.

the body. In this animal what were thought to be the origins of the left and right aortae were dilated. At autopsy the aortae and origins of the great vessels showed evidence of dissecting aneurysms (Fig. 7).

Abscesses

Palpable or visible swellings are common veterinary diagnostic problems. Some idea of the solidity or fluidity of their contents may assist management decisions. For instance a grey rat snake thought to have a neoplasm was examined by both X-ray and ultrasound. The former showed a homogeneous swelling, but the latter revealed the partly fluid centre. It proved to be an abscess formed around a piece of bone that had penetrated the gut wall. After surgical treatment the snake recovered.

Discussion

There are numerous practical advantages in having a non-invasive, non-traumatic method of diagnosing and monitoring pregnancy in zoo animals.

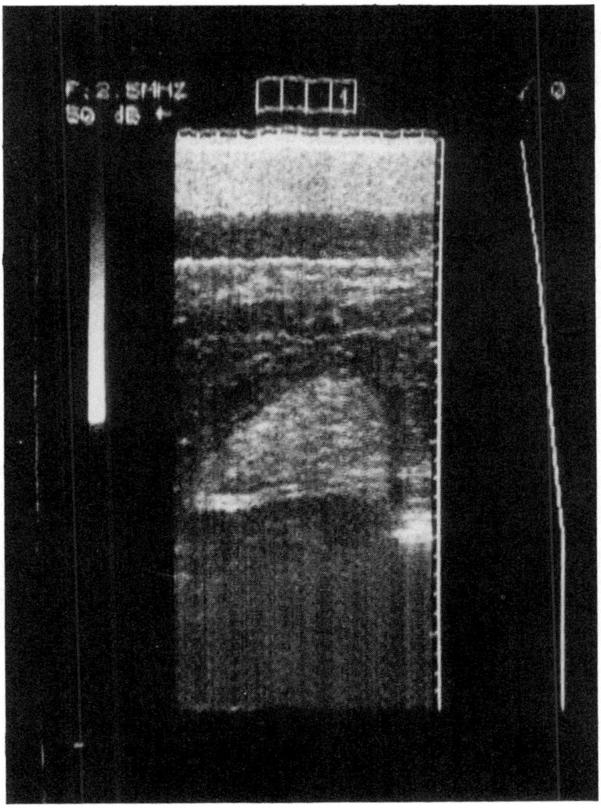

Fig. 5. Killer whale: abnormal liver. A straight linear array probe was used, as in Fig. 3. In this case the probe produced sound at 2.5 MHz, providing greater penetration. Beneath the successive layers of skin, blubber, subcutaneous tissue and muscle, the liver has been visualized. Within this darker grey structure there is a well-defined region of brighter signal. This represents fatty degeneration, probably due to malnutrition.

Early certainty about the success or failure of mating can improve the economy of management. It can also provide information that may advance the understanding of hormone assay results and throw light on the behaviour of animals whose reproductive physiology is not well understood. Some of these managemental advantages were seen in the blackbuck. Once diagnosis and monitoring of pregnancy could be carried out by ultrasound instead of by urine sampling, it was no longer necessary to confine the animal in a restricted area for long periods. Another instance of the value of the technique in animal management was seen when ultrasonic investigation resolved doubt about which of two pregnant female chimpanzees had given birth to an infant being carried by one of them.

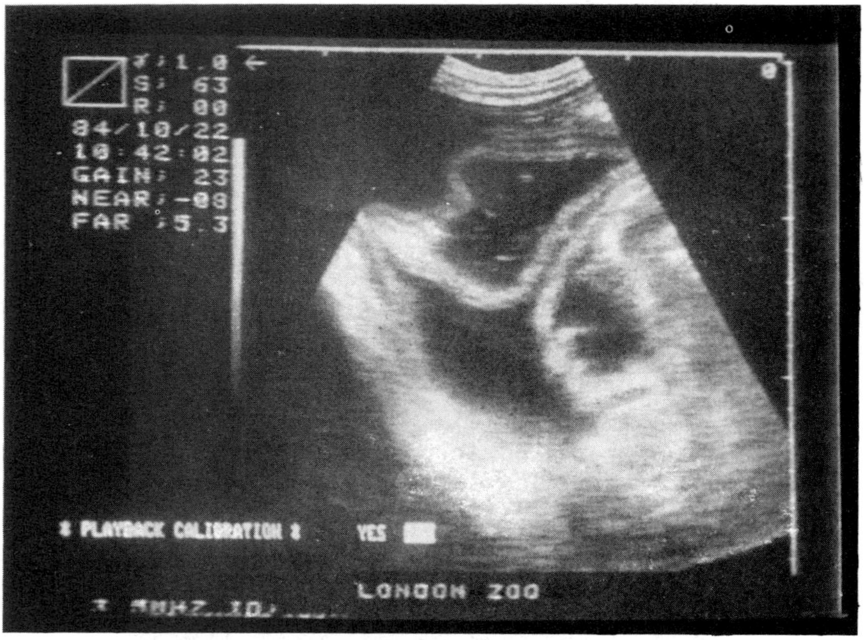

Fig. 6. Giant panda. A 3.5 MHz curved linear array probe was used. A small segment at the top of the picture represents the abdominal wall. Immediately beneath this, at the top left of the picture, there is a dark region caused by the presence of ascitic fluid. A second area of ascites lies in the lower central part of the picture. Between these two areas lies a cross section of a dilated loop of gut, containing fluid and faecal debris and having an irregular thickened wall.

Late prepartum diagnosis of foetal position may also prove valuable in some species. Cho, Hanari, Suzuki & Honjo (1985) were able to relate a high stillbirth rate in cynomolgus monkeys to foetal position prior to labour.

Unlike X-rays ultrasound does not require immobilization for most clinical purposes, and neither does it present any radiation risk to the animal or the helpers. Some animals, however, are not tame enough to be examined without sedation or anaesthesia, and the possible hazards of such procedures then have to be weighed against the advantages of the information to be obtained. So sedation of the rhesus monkeys examined by Michejda & Hodgen (1981) and Sabbagha, Turner & Chez (1975), who used a B-scanner, was probably necessary because the monkeys were not tame enough to examine otherwise, but thought justifiable because the examination made it possible to obtain a range of biparietal diameters.

To make the most use of the potential of ultrasound in this field, an intrarectal or vaginal approach may be needed, at least in ungulates. For medium- or small-sized animals this requires a small-parts probe. Much,

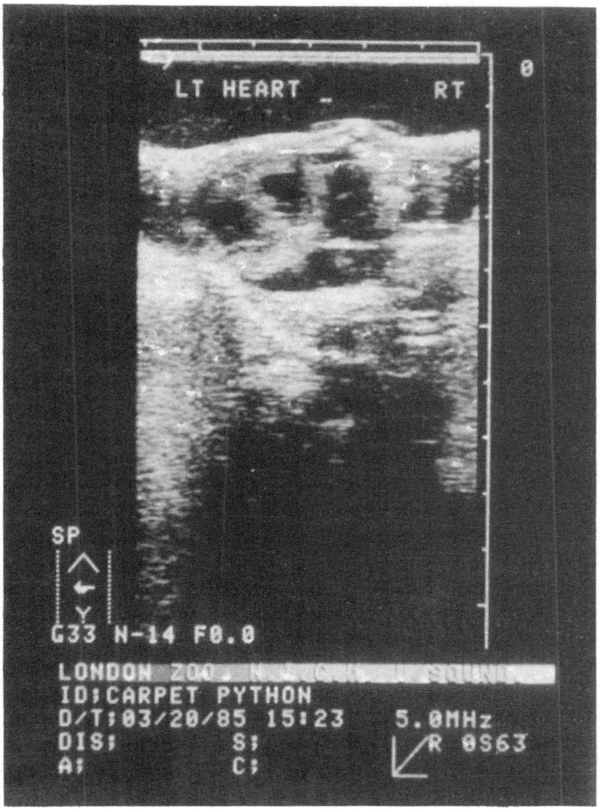

Fig. 7. Carpet python: heart and great vessels. A 5 MHz straight linear array with a water-path standoff was used. The black area at the top represents the water-path. It is only the central part of the image, between the left and right markers, that constitutes the cross section through the snake's heart. The top of the picture is the ventral surface of the snake. On the left, two dark thick-walled cavities are the dilated conjoined ventricle, seen at the level of the incomplete interventricular septum. To the right and dorsal to this is a curved vessel (black, crossed by a white reverberation artefact). This was a dissecting aneurysm of the aorta.

however, can be achieved by external examination. Whichever route is employed, the skills of examination and interpretation can only be learnt after considerable practice.

Being able to pinpoint some of the causes of failure to breed, such as disease of the uterus or adnexa, may be as great an advantage as the ability to monitor pregnancy.

The ability to recognize the existence of a twin at an early stage may be helpful. In breeding racehorses, for example, it may enable the selective ablation of one gestation sac to be carried out. Indeed, ultrasound may be an essential technique for the operation itself (J. R. Newcombe, pers. comm.).

For disease diagnosis ultrasound is still at the very beginning of its development in veterinary practice. Because of its safety, flexibility and ability to differentiate between tissues it holds as much promise as X-radiology, but the skills required are learnt only by a great deal of application and need to be concentrated in the hands of the person who is going to be the specialist.

References

Cho, F., Hanari, K., Suzuki, M.T. & Honjo, S. (1985). Relationship between fetal position and stillbirth in the Cynomolgus monkeys (*Macaca fascicularis*): retrospective analysis. *J. med. Primatol.* **14**: 169–74.

James, A.E., Brayton, J.B., Novak, G., Wight, D., Shehan, T.K., Bush, R.M. & Sanders, R.C. (1976). The use of diagnostic ultrasound in evaluation of the abdomen in primates with emphasis on the Rhesus monkey (*Macaca mulatta*). *J. med. Primatol.* **5**: 160–75.

Michejda, M. & Hodgen, G. (1981). *In utero* diagnosis and treatment of non-human primate fetal skeletal abnormalities. *J. Am. med. Ass.* **246**: 1093–7.

Nyland, T.G., Hill, D.E., Hendrickx, A.G., Farver, T.B., McGahan, J.P., Henrickson, R., Anderson, J. & Phillips, H.E. (1984). Ultrasonic assessment of fetal growth in the non-human primate (*Macaca mulatta*). *J. clin. Ultrasound* **12**: 387–95.

Sabbagha, R.E., Turner, J.E. & Chez, R.A. (1975). Sonar biparietal diameter growth standards in the rhesus monkey. *Am. J.Obst. Gynec.* **121**: 371–4.

Yeager, C.H., O'Grady, J.P., Estra, G., Thomas, W., Kramer, L. & Gardner, H. (1981). Ultrasonic estimation of gestational age in the lowland gorilla: a biparietal diameter growth curve. *J. Am. vet. med. Ass.* **179**: 1309–10.

Symp. zool. Soc. Lond. (1988) No. 60 151–158

Immunoassays with special reference to ELISA methods

A. VOLLER and D. E. BIDWELL

Nuffield Laboratories of Comparative Medicine
Institute of Zoology
The Zoological Society of London
Regent's Park
London NW1 4RY, UK

Synopsis

Radioimmunoassay was established for hormone measurement and immunofluorescence for the diagnosis of infectious diseases. The Enzyme-Linked Immunosorbent Assay combines the merits of these two methods and has been widely applied throughout diagnostic medicine. Monoclonal antibodies and novel test formats have led to the development of tests suitable for use outside conventional laboratories.

Introduction

Laboratory methods have become essential in the diagnosis of virtually all types of diseases. Recently there has been particular emphasis on the use of immunological methods, generally known as immunoassays. Our unit has been deeply committed to the study of such methods at the Zoological Society of London for over two decades. During this time we have participated in a wide range of immunodiagnostic studies on both humans and animals. Some of the developments and applications of this work will be reviewed in this paper.

Immunoassays

Immunoassays are based on immune reactions for the measurement of biological substances. The simplest assays were based on the precipitation or agglutination of antigens or antibodies. Such methods led to the understanding of the red cell antigen-antibody systems and blood transfusion is still heavily dependent on them. However, they do not have adequate sensitivity for the assay of many infectious agents, cell constituents and hormones.

ZOOLOGICAL SYMPOSIUM No. 60
ISBN 0-19-854007-8

Radioimmunoassay

The need for more sensitive methods led to the concept of labelled reagent immunoassays. In this, one immuno-partner (antibody or antigen) is marked with a label which can be detected and measured at very low concentrations. Initially the label was an isotope (Yallow & Berson 1960) and this was the basis of radioimmunoassay (RIA) which effectively founded modern endocrinology and led to a Nobel prize for the originators. These assays were all based on the premise that hormone in a sample would compete with the provided labelled hormone for the opportunity to combine with a limited amount of antibody, e.g.

$$\mathrm{Ag}^* + \mathrm{Ab} \rightarrow \mathrm{AbAg}^*$$

$$(\mathrm{Ag}^* + \mathrm{Ag}) + \mathrm{Ab} \rightarrow \mathrm{AbAg} + \mathrm{Ag}^*$$

$$(^* = \text{isotopic label})$$

Thus these were all 'competitive' or 'limited labelled-reagent assays'. The more antigen that there was in the sample the less the labelled antigen could combine with the antibody. Either the remaining free labelled antigen or the labelled antigen that had combined with the antibody could be measured accurately at very low concentrations. These methods are still the basis for virtually all endocrinological measurements.

In spite of the virtues of RIA, certain drawbacks were recognized. The isotopic label decayed and reagents had only a short usable life. The labelling was restricted to levels beyond which the immunological reactivity of the reagents was reduced.

Virtually all the original RIA work was carried out in clinical chemistry laboratories and the science of endocrinology was in effect based on the ability to measure low concentrations of hormones by RIA.

Immunofluorescence

However, a different approach was being developed for infectious diseases. In this field there was a need to detect both antibodies and antigens at very low concentrations. RIA was often unsuitable for this purpose and medical and veterinary laboratories turned to the use of non-isotopic labels, in particular fluorescent labels, in a group of techniques known as immunofluorescence. The exact method for detecting and measuring antibody was as outlined below:

$$|\ \mathrm{Ag} + \mathrm{Ab} + \text{anti Ig}^{\mathrm{F}} \rightarrow \quad |\ (\mathrm{AgAb}\ \text{anti Ig}^{\mathrm{F}})$$

The infectious agent was immobilized on a solid surface, usually a microscope slide. It was then made to react with the patient's sample being examined for antibody, and later with an antispecies immunoglobulin labelled with fluorescein. The result of the reaction was visualized by fluorescence microscopy. The amount of antibody could be estimated roughly by observing the intensity of fluorescence with different dilutions of the test sample.

This method was in fact used for the measurement of antibody in virtually every infectious disease of man and animal as well as in many auto-immune diseases. At the Zoological Society of London we developed immunofluorescent methods which were especially applicable to the immunodiagnosis of parasitic diseases in man and animals (Voller & Houba 1980). We were able to assist in the establishment of these methods throughout the tropical world with the assistance of the Medical Research Council of the UK, The World Health Organization and The Wellcome Trust. Even now the indirect immunofluorescence method is considered a reference test for African trypanosomiasis, malaria, amoebiasis, Chagas disease, and many other infections.

This method was particularly attractive to infectious disease laboratories because by the use of class-specific anti-immunoglobulin reagents (e.g anti IgM) the sub-class of the antibody could be measured. This information was sometimes of value in indicating whether infection was recent or long past (e.g. in toxoplasmosis).

Enzyme immunoassay

In 1968 Miles & Hales suggested that non-isotopic labels might be used in place of isotopes and this was followed after a few years by the publication of the first enzyme immunoassay methods (Engvall & Perlmann 1971) which will occupy the major part of this paper.

Enzyme immunoassay is another non-isotopic labelled-reagent method which combines the merits of radioimmunoassay and immunofluorescence but which avoids many of the disadvantages of those methods. In enzyme immunoassay an enzyme is used to label either antibody or antigen in such a manner that immunological activity is retained. The enzyme (and hence its immuno-carrier) can be detected at very low concentrations by the addition of a substrate capable of yielding a detectable signal in the form of colour, fluorescence, luminescence etc.

Some of the earliest enzyme immunoassays were directly analogous to RIA, merely replacing the isotopic label with the enzyme. Several assays for hormones were established in this way (Van Weemen & Schuurs 1971).

(a) $Ag^E + Ab \rightarrow AbAg^E$ (no antigen in sample)

(b) $(Ag^E + Ag) + Ab \rightarrow AbAg + Ag^E$ (antigen in sample)

Indeed, these methods are still in use and have proved convenient and reliable for the measurement of many haptens. Progesterone is a good example and the test is used as an indicator of pregnancy in wild and domestic animals.

Earlier, Miles & Hales (1968) had suggested that non-competitive immunoassays for antigens could be advantageous. The test outline they suggested was of the sandwich or immunometric type, e.g.

$$\mathrm{Ab} + (\mathrm{Ag}) + \mathrm{Ab}^{\mathrm{I}^{125}} \rightarrow \mathrm{AbAgAb}^{\mathrm{I}^{125}}$$

This method had the advantage that the amount of analyte present in the sample was reflected in the strength of signal generated. However, the major advance consisted in combining a solid phase system and a non-isotopic labelled reagent to produce an immunometric system. This concept was first formulated in Sweden by Engvall & Perlmann (1971, 1972) and was named ELISA (enzyme-linked immunosorbent assay). The method could be used for antibody measurement as follows:

Step 1. $| \mathrm{Ag} + \mathrm{Ab} \rightarrow \quad | \mathrm{AgAb}$

Step 2. $| \mathrm{AgAb} + \text{anti } \mathrm{Ab}^{\mathrm{E}} \rightarrow \quad | (\mathrm{AgAb} \text{ anti } \mathrm{Ab}^{\mathrm{E}}) + \text{substrate} \rightarrow \text{colour}$

The antigen was immobilized on a plastic tube by simple adsorption. The coated tube was then allowed to react with the test sample containing antibody. After capture of the antibody by the antigen, other constituents of the sample were washed away. The captured antibody was then indicated by the addition first of enzyme-labelled antiglobulin and then of enzyme substrate.

This may sound complicated but in reality provides an extraordinarily simple and sensitive method for universal application. Virtually all antigens of protein, lipoprotein or nucleo-protein composition will become fixed to plastic by adsorption, and are usually stable. The coated plastic can be moulded into almost any form and permits very easy washing. The sample is tested at only a single dilution, the conjugate is stable, and the substrate reaction can be read either by eye or, to abolish subjectivity, in a simple colorimeter.

A technical development of ELISA consisted in the use of plastic microplates to carry the immobilized antigen or antibody. This advance, permitting the easy manipulation of large numbers of test samples and reagents, was made at the Zoological Society of London and has since become the standard for virtually all ELISA tests (Voller, Bartlett, Bidwell, Clark & Adams 1976). Our group was privileged to introduce the microplate ELISA for the measurement of antibody in numerous infectious diseases including malaria, sleeping sickness, toxoplasmosis, rubella, cytomegalovirus infection and measles, to name but a few (Voller, Bidwell & Bartlett 1979). Other workers used similar approaches to diagnose many viral, bacterial and parasitic diseases in man and in domestic and wild animals (Wardley &

Crowther 1982; Avrameas, Druet, Masseyeff & Feldmann 1983). More recently the microplate ELISA for antibody has found wide application in the screening for antibody to human immunodeficiency virus (the causative agent of AIDS).

At an early stage it was appreciated that an immunometric type ELISA could be used to detect and measure antigen by means of enzyme-labelled specific antibody, e.g.

Step 1. | Ab + Ag → | AbAg

Step 2. | AbAg + Ab^E → | $AbAgAb^E$+ substrate → colour

In this procedure specific antibody was adsorbed to the plastic surface. The coated plastic was then allowed to react with the sample, any antigen being captured and other components washed away. The subsequent addition of labelled antibody completed the 'sandwich' and imparted high specificity to the system.

This method was introduced in 1976 for the detection of hepatitis B virus surface antigen (HBsAg) in blood (Wolters, Kuijpers, Kacaki & Schuurs 1976) and for pathogenic plant viruses (Voller, Bidwell & Bartlett 1976). Since that time the 'double antibody sandwich' ELISA, as it is often called, has become the test of choice for HBsAg, rotaviruses, arboviruses and many other human pathogenic agents. Our group at the Zoological Society of London were gratified to see that their initial work in the plant virus field has led to the ELISA method becoming firmly established as the preferred method for monitoring the viral status of vegetables, fruit trees, flowers and seed throughout the world. There are even kits so that farmers and plant pathologists can diagnose 50 or more agriculturally important viral disorders.

The merit of the double antibody sandwich ELISA is that tiny amounts of virtually any macromolecule can be detected and measured in almost any kind of sample without interference from the sample. This is because the first step of the test system is essentially an immunopurification procedure. Thus a high sensitivity can be obtained for a particular antigen even in the presence of large amounts of contaminatory material. For example, it was found that, as in isotopic assays, hepatitis B surface antigen could be measured in blood at concentrations of about 1 ng/ml, and other pathogens (e.g. rotaviruses) could be detected in faecal samples without any pretreatment of the samples. The same advantages also applied to the detection of antigens in malignant and other non-infectious diseases. The method proved to be ideal for the measurement of alpha-foetal protein (AFP) to diagnose hepatocellular carcinoma and for pre-natal screening for AFP as an indicator of neural tube defects (Masseyeff, Maiolini, Ferrua & Raginbeau-Gilli 1976; Barlow, Thompson, Cuckle, Wald & Voller 1986). There is an ever increasing list of the applications of the antigen ELISA method; these range from the measurement *in vivo* of therapeutic monoclonal antibodies, to the assay of soya protein in processed foods (Hitchcock, Bailey, Crimes, Dean & Davis 1981), to the identification of mosquito blood meals (Service, Voller & Bidwell 1986).

Newer developments

Hybridoma techniques for the production of antibodies *in vitro* were developed at the same time as enzyme-immunoassays and this led to a natural marriage of the two, resulting in many monoclonal antibody-based ELISA tests. The advantage of the high specificity of monoclonal antibodies was coupled with that of the unlimited quantities in which such antibodies were available.

One particular ELISA configuration for antigen detection has led to considerable dispute and litigation. This is known as the two-site monoclonal test. It is a sandwich ELISA but the monoclonal antibody on the solid phase plastic is directed against one epitope of the antigen and a second monoclonal, which is enzyme-labelled, is reactive with a different epitope of the antigen.

$$| \mathrm{McAb}^1 + (\mathrm{Ag}) + \mathrm{McAb}^{2^E} \rightarrow \quad | \mathrm{McAb}^1\mathrm{AgMcAb}^{2^E}$$

The virtue of this method is that the complete test can be reduced to a single incubation step.

Traditionally ELISA results are based on a change in colour of a substrate solution. Improvements are constantly being made on this aspect of the test. Recently an amplified substrate system has been developed whereby the colour generation is increased several hundred times by a substrate cascade method (Self 1985). Another variation on this theme is the choice of a substrate that yields a luminescence signal on its degradation (Whitehead, Thorpe, Carter, Groucutt & Kricka 1983). Commercial manufacturers make much of these minor modifications, but they can be of special value in particular circumstances.

There is now a trend towards the performance of diagnostic tests outside the conventional laboratory. ELISA has played a major role here and is already represented by home pregnancy tests, ovulation prediction kits, and many agricultural and veterinary products. There is also increasing use of ELISA in the measurement of drugs and other small molecules (haptens) for both clinical and other purposes. We anticipate that ELISA-type methods will continue to expand into new areas including dipsticks, test-strips etc. to bring the advantages of immunoassays to an ever-widening circle of potential users.

Acknowledgements

We are grateful to the Zoological Society of London for providing facilities and assistance for our studies over the years and to the Medical Research Council (UK), World Health Organization and Wellcome Trust for financial support.

References

Avrameas, S., Druet, P., Masseyeff, R. & Feldmann, G. (Eds). (1983). *Immunoenzymatic techniques*. Elsevier, Amsterdam.

Barlow, R.D., Thompson, S.G., Cuckle, H.S., Wald, N.J. & Voller, A. (1986). Comparison of ELISA with an RIA for AFP determination in screening for neural tube defects. *Ann. clin. Biochem.* **23:** 334–9.

Engvall, E. & Perlmann, P. (1971). Enzyme-linked immunosorbent assay. *Immunochemistry* **8:** 871–4.

Engvall, E. & Perlmann, P. (1972). ELISA III, quantitation of specific antibodies by enzyme labelled antiglobulin in antigen coated tubes. *J. Immunol.* **109:** 129–35.

Hitchcock, C.H.S., Bailey, F.J., Crimes, A.A., Dean, D.A.G. & Davis, P.J. (1981). Determination of soya in food using ELISA. *J. Sci. Food Agric.* **32:** 157–65.

Masseyeff, R., Maiolini, R., Ferrua, B. & Raginbeau-Gilli, J. (1976). Quantitation of AFP by enzyme immunoassay. In *Protides of the biological fluids* **24:** 605–12. (Ed. Peeters, H.) Pergamon Press, Oxford.

Miles, L.E.M. & Hales, C.N. (1968). Labelled antibodies and immunological assay systems. *Nature, Lond.* **219:** 186–9.

Self, C.H. (1985). Enzyme amplification. A good method applied to provide an immunoassisted assay for placental alkaline phosphatase. *J. immunol. Meth.* **76:** 389–94.

Service, M.W., Voller, A. & Bidwell, D.E. (1986). The ELISA test for the identification of blood meals of haematophagous insects. *Bull. ent. Res.* **76:** 321–30.

Van Weemen, B.K. & Schuurs, A.H.W.M. (1971). Immunoassay using antigen-enzyme conjugates. *FEBS Lett.* **15:** 232–4.

Voller, A., Bartlett, A., Bidwell, D.E., Clark, M.F. & Adams, A.N. (1976). The detection of viruses by ELISA. *J. gen. Virol.* **33:** 165–7.

Voller, A., Bidwell, D.E. & Bartlett, A. (1976). Enzyme immunoassays in diagnostic medicine: theory and practice. *Bull. Wld Hlth Org.* **53:** 55–65.

Voller, A., Bidwell, D.E. & Bartlett, A. (1979). *The enzyme-linked immunosorbent assay—ELISA: a guide with abstracts of microplate applications*. Dynatech Europe, Guernsey.

Voller, A. & Houba, V. (1980). Malaria. In *Immunological investigation of tropical parasitic diseases*: 6–31. (Ed. Houba, V.). Churchill Livingstone, Edinburgh.

Wardley, R.C. & Crowther, J.R. (Eds). (1982). *The ELISA in veterinary research and diagnosis*. Martinus Nijhoff, The Hague.

Whitehead, T.P., Thorpe, G.H.C., Carter, T.J.N., Groucutt, C. & Kricka, L.J. (1983). Enhanced luminescence procedure for sensitive determination of peroxidase labelled conjugates in immunoassay. *Nature, Lond.* **305:** 158–9.

Wolters, G., Kuijpers, L., Kacaki, J. & Schuurs, A.H.W.M. (1976). Solid phase enzyme immunoassay for detection of HBsAg. *J. clin. Path.* **28**: 873–6.

Yallow, R.S. & Berson, S.A. (1960). Immunoassay of endogenous plasma insulin in man. *J. clin. Invest.* **39**: 1157–75.

Symp. zool. Soc. Lond. (1988) No. 60 159–173

Anaerobic bacteria as pathogens in wild and captive animals

G. R. SMITH

Nuffield Laboratories of Comparative Medicine
Institute of Zoology
The Zoological Society of London
Regent's Park
London NW1 4RY, UK

Synopsis

Necrobacillosis is caused by *Fusobacterium necrophorum*, a normal inhabitant of the gut of herbivores and other animals. Its host range includes domesticated and wild animals, and man. Organisms from the faeces invade damaged epithelium and produce, probably by means of various toxins and exoenzymes, necrotizing infections of the mouth, foot, liver and other sites. Synergistic interactions with other bacteria sometimes occur. Infection results in antibody production but little if any immunity, and prospects for a satisfactory vaccine are poor.

The disease occurs in captive and free-living macropods, lesions of the mouth, face and stomach wall being especially common. The Macropodidae do not have a uniquely low resistance to *F. necrophorum* but are often exposed to certain important predisposing factors. Control of the disease in captive macropods is of necessity by management rather than by immunization.

Necrobacillosis in free-living deer and antelope produces heavy losses in parts of North America, Asia and Africa. The lesions include stomatitis, foot-rot and rumenitis. Captive animals are also affected, particularly under conditions of poor hygiene.

Botulism, once thought to be invariably a pure intoxication due to ingestion of preformed toxin, can occasionally be caused by toxigenesis in the gut—as in human 'infant botulism'. This requires a favourable intestinal microflora. Analogous microbiological factors may exert an effect on *C. botulinum* in the environment, e.g., in muds that inhibit the organism, and also in carrion—probably the main source of toxin for animals.

The real incidence of botulism in zoos is probably much higher than that reflected in recorded outbreaks. The disease in captive animals arises from the consumption of spoiled foods, contaminated maggots, rotting carcasses of birds that have died after ingesting toxin in the natural environment, or toxic organic material from slow-moving water. Huge losses occur annually in free-living waterfowl in temperate climates. The status of botulism in waterfowl and wild mammals in the tropics is unknown and requires investigation.

ZOOLOGICAL SYMPOSIUM No. 60
ISBN 0-19-854007-8

Introduction

By comparison with man, animals live in close contact with soil and with their own faeces. They also often relish spoiled foodstuffs, including rotting carcasses, that would be repugnant to man. As a result two pathogenic anaerobes—*Fusobacterium necrophorum* and *Clostridium botulinum*—have assumed a place of particular importance in the causation of animal disease. *F. necrophorum*, being a normal inhabitant of the gut, is frequently present in faeces; and *C. botulinum*, whose natural habitat is soil and mud, produces in rotting organic matter one of the most potent of all the biological toxins. Both organisms cause significant loss in free-living and captive animals.

Necrobacillosis

Necrobacillosis is a necrotizing infection caused by *Fusobacterium necrophorum*, sometimes by itself but often mixed with other bacteria—anaerobes, facultative anaerobes, and aerobes. It occurs in a wide range of domesticated and wild animals, and occasionally in man.

Aetiological agent

The organism now known as *F. necrophorum* was first described in 1884 by Loeffler, who recognized it as the cause of calf diphtheria. This gram-negative non-sporing anaerobe belongs to the Bacteroidaceae. It varies in morphology from coccobacilli to unbranched filaments up to 100 μm in length. Staining is often irregular, giving rise to beaded forms. The cells are oxygen-sensitive, and liquid cultures are killed within 30 min by a temperature of 56°C. The organism is susceptible to many antibiotics including penicillin. Many strains isolated from cases of disease produce a lipopolysaccharide endotoxin, at least one exotoxin (a leucocidin), a haemolysin, a haemagglutinin, a lipase, and a DNAase. Some of these factors may play a role in pathogenicity. So-called phase A strains (Beerens, Fievez & Wattre 1971; Hofstad 1985) are haemolytic, haemagglutinating, highly leucocidino genic, and pathogenic for mice; phase B strains produce less leucocidin, are haemolytic but not haemagglutinating, and have no more than slight pathogenicity for mice; phase C strains lack haemolytic, haemagglutinating and pathogenic properties.

Habitat of *Fusobacterium necrophorum*

The organism occurs as a commensal in the alimentary tract of herbivores and other species including pigs and dogs. For example, at least 50% of cattle carry it in the rumen in concentrations of 10^3–10^6 viable organisms per gram of ruminal contents. In this situation organisms in phase B are more common than those in phase A. *F. necrophorum* is voided in the faeces, in which it remains viable for limited periods, but multiplication seldom occurs on the ground. Necrobacillosis lesions commonly yield phase A organisms.

Pathogenesis

Infection is usually endogenous, the source of organisms being the faeces. Filthy surroundings, particularly those with large accumulations of dung underfoot, are conducive to the disease. *F. necrophorum* has little ability to invade intact epithelium, but when epithelial damage occurs, for example as a result of trauma, maceration or infection (bacterial or viral), the organism enters the tissues, multiplies profusely and—probably assisted by some of its toxins and exoenzymes—produces coagulative necrosis and abscess formation. Histologically, the picture is one of a structureless area of necrosis surrounded by a mass of filamentous bacteria and a zone of leucocytes. The infection may be pure but if mixed is often accompanied by a putrid smell. Lesions of the external body surface or mucosal surfaces frequently become suppurative and gangrenous in the later stages. Lesions—often though not always yielding *F. necrophorum* in pure culture—also occur in the viscera, especially the liver and lung, usually as the result of haematogenous spread of bacteria from infected foci of the body surface or mucosae.

Necrobacillosis can affect almost any part of the body, but the three sites in which it occurs most commonly are the mouth, liver and foot. Oral necrobacillosis, well exemplified by calf diphtheria, sometimes occurs simultaneously in the majority of members of a group as a result of predisposing factors affecting them all, rather than of contagion. Necrobacillosis of the liver affects numerous species but is particularly well known in 'feedlot cattle', in which the sudden change from high-fibre to high-concentrate diet may lead to ruminal acidity and rumenitis. *F. necrophorum* may then infect the epithelial lesions and subsequently find its way via the portal circulation to the liver (Jensen, Deane, Cooper, Miller & Graham 1954). In foot rot of cattle and many other ungulates the primary agent is *F. necrophorum*, which invades tissue damaged for example by hard rubble underfoot or by excessively muddy surroundings. In severe cases the infection spreads to the interphalangeal joints. Sheep foot rot, it should be noted, is a truly contagious disease caused mainly by *Bacteroides nodosus*, but *F. necrophorum* is also concerned.

Some of the numerous species of bacteria that frequently accompany *F. necrophorum* in lesions of necrobacillosis may act synergistically. For example the leucocidin of *F. necrophorum* assists the establishment of *Corynebacterium pyogenes* in the tissues, and *C. pyogenes* produces a substance that stimulates the growth of *F. necrophorum* (Roberts 1969).

Experimentally, subcutaneous infection of mice with virulent strains of *F. necrophorum* requires the injection of several million organisms, though much smaller doses are sufficient by the intracerebral route (Smith, Oliphant & Parsons 1984). It was shown by Smith & Turner (1986) that dilution had an adverse effect on the infectivity of 24-h culture, which became apparent at or near the minimum lethal dose. Thus in mice inoculated subcutaneously the mortality produced by 0.01 ml of undiluted culture was almost invariably greater than that produced by 0.1 ml of a 1 in 10 dilution. The explanation appeared to lie in the increased physical separation of bacterial cells.

Immunity

Infection with *F. necrophorum* gives rise to antibodies detectable by serological techniques (Warner, Fales & Teresa 1974; Evans & Berg 1985; Turner, Bidwell & Smith 1986), but there has been much uncertainty as to whether increased resistance results from infection and from vaccination. Simon & Stovell (1969) reviewed the conflicting opinions contained in the literature from 1891 to 1960 but commented that the frequent presence of *F. necrophorum* abscesses of different ages within a single bovine liver indicated that natural infection did not confer immunity. Smith, Turner, Murray & Oliphant (1985) referred to the more recent literature, which contained several encouraging reports.

Against this background of uncertainty Smith, Turner, Murray *et al.* (1985) carried out experiments in mice. Two extreme methods of immunization with killed vaccines gave no trace of protection. Also unsuccessful as a means of immunization was the production of a severe and potentially lethal subcutaneous infection with *F. necrophorum*, followed by curative treatment with metronidazole. Slight but significant protection against subcutaneous challenge resulted, however, from two such infections given in rapid succession. It was concluded that the main virulence factors of *F. necrophorum* were only weakly immunogenic, and that the experiments gave little encouragement to the prospect of an effective necrobacillosis vaccine.

Necrobacillosis in macropods

'Kangaroo disease' or 'lumpy jaw' is well known in all parts of the world as the commonest affliction of macropods in captivity, and numerous descriptions have been published since the early years of this century. The sites of the body most often affected are the periodontal tissues and face, the stomach wall, hind limbs, liver and lung. Fox (1923: 570–95) gave a detailed account of cases at the Zoological Society of Philadelphia. Scott (1925) in reviewing the period 1909 to 1924 at the Zoological Society of London, concluded that 61 (34%) of 180 deaths in marsupials could be attributed to 'streptothricosis' (necrobacillosis). He reported that no fewer than 14 of a group of 19 Bennett's wallabies (*Macropus bennetti*) died from the infection within four months of being introduced into the Collection in December 1924. Over half a century later, in the first four months of 1979, more than 200 deaths from the disease occurred in a population of *c.* 700 red-necked wallabies (*Macropus rufogriseus*) running free at Whipsnade (Ashton 1982: 91–2). The stomach wall and face were the most commonly affected sites. Because the weather at the time was severely cold a daily food supplement was given to make good the curtailed natural feeding. The daily gathering of animals at one site may have contributed to the exceptional severity of the outbreak.

The disease also occurs in wild macropods. Typical cases in kangaroos—mainly the red kangaroo (*Macropus rufus*)—in the Murchison area and elsewhere in Western Australia were described by Tomlinson & Gooding (1954). Information available to them suggested that outbreaks had occurred frequently for at least the preceding 50 years.

Aetiology

The disease in macropods is well known, but its aetiology has been the subject of much debate, having been variously ascribed to *Nocardia* spp., *Actinomyces* spp., *Fusobacterium necrophorum*, *Bacteroides* spp., *Bacteroides ruminicola* subsp. *brevis*, and other organisms (Fox 1923; Beveridge 1934; Watts & McLean 1956; Finnie 1976; Keane, Taylor, Wilson, Smith, Cunningham, Devine & English 1977; Burton 1981).

The disease as seen at the Zoological Society of London was the subject of a detailed study by Oliphant, Parsons & Smith (1984). Their findings (Fig. 1) showed that *F. necrophorum* was the main aetiological agent, being present in 69% of lesions, usually in heavy and occasionally in pure culture. *Bacteroides* spp., six in all, often in considerable numbers, were found in 48% of lesions, and *Fusobacterium nucleatum* in 14%. Aerobes or facultative anaerobes, usually in small numbers, were present in 59% of lesions. The organisms isolated often varied considerably from one part of a lesion to another. This fact, together with the need for great care to exclude contaminants during sampling and for impeccable anaerobic technique, may have been the cause of much of the controversy in the past. In Australia, Samuel (1983), like Oliphant *et al.* (1984), concluded that of all the organisms isolated from the lesions *F. necrophorum* was the main pathogen.

The factors that predispose macropods to infection of the mouth and face by damaging the oral mucosa include unusual dentition (see Tyndale-Biscoe 1973) and the presence of sharp penetrating material such as cereal awns in food. Extreme agility predisposes to cuts and abrasions of all parts of the external body surface, particularly when the animals are crowded together. In addition, prolonged overcrowding within a limited area leads to the

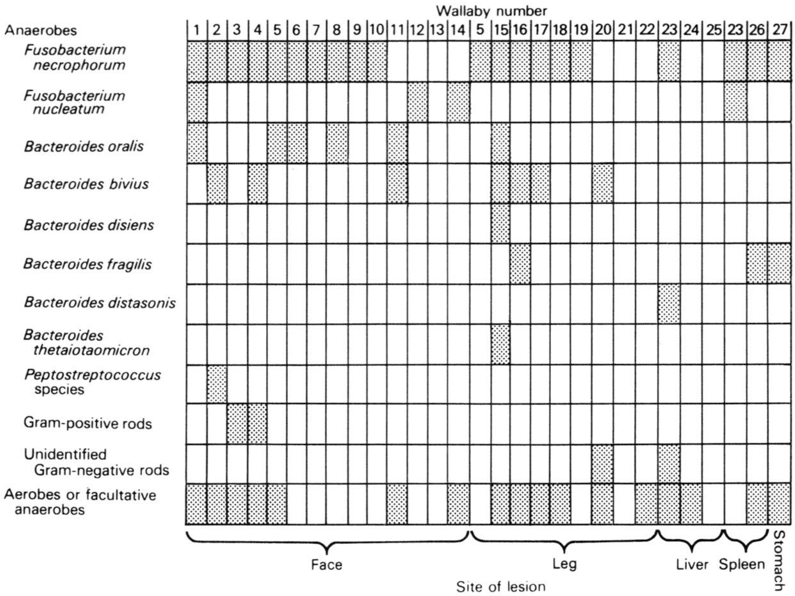

Fig. 1. Bacteria isolated from lesions of necrobacillosis in 27 wallabies. Wallabies 5 and 23 were each affected in two sites. ■ Positive culture. □ Negative culture. From Oliphant *et al.* (1984) with permission of *Research in Veterinary Science.*

build-up of faeces on the ground, with concomitant exposure of animals to *F. necrophorum.*

Few would disagree with Finnie's (1976) comment that 'macropods in captivity are very susceptible' to necrobacillosis. The astonishing degree of susceptibility is, however, probably related to the predisposing factors mentioned above rather than to a uniquely low resistance to *F. necrophorum* infection. Thus Smith, Turner & Cinderey (1986) found that wallabies were not appreciably more susceptible than rabbits or mice to *F. necrophorum*—a fact established by the subcutaneous injection of a series of graded doses simultaneously into animals of each species.

Pathology

Facial and gastric necrobacillosis in macropods require special comment. The facial lesions usually arise from infections of the mouth, particularly the periodontal tissue. The face becomes swollen and the infection may spread to the orbit and jawbones, in which a rarefying osteitis occurs (Fig. 2). Exostoses may form in protracted infections. Gastric lesions are often very severe and may lead to perforation and peritonitis. Infections are accompanied by increased fibrinogen, abnormal neutrophil morphology, and often by neutrophilia (Hawkey, Ashton, Hart, Cinderey & Jones 1982).

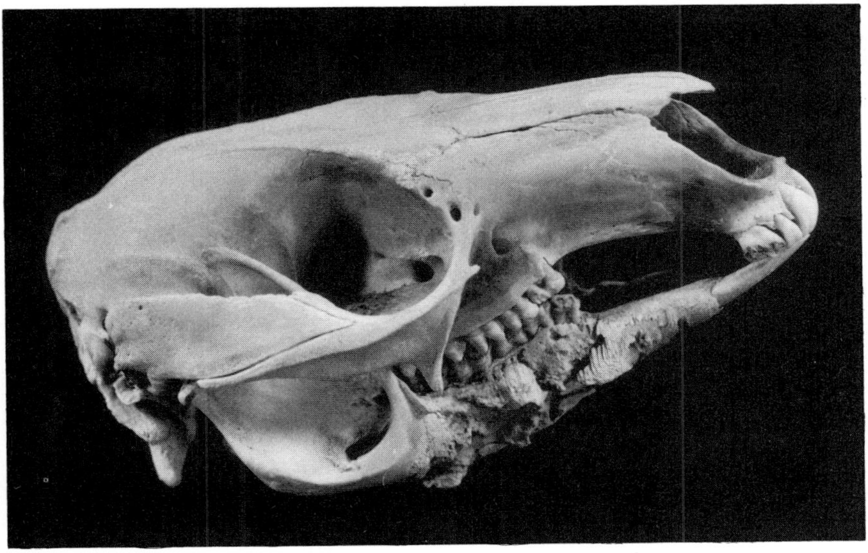

Fig. 2. Erosion of mandible in a wallaby with necrobacillosis.

Immunity and control

Smith, Turner & Cinderey (1986) attempted to immunize wallabies by means of a vaccine consisting of a large dose of heat-killed *F. necrophorum* culture emulsified with a powerful experimental adjuvant (Freund's complete). There was no suggestion of increased resistance when the animals were challenged subcutaneously eight weeks later with a dose of *F. necrophorum* considered to be moderate in size. The experiment, although small in scale, supported the results of the extensive investigation made earlier in mice by Smith, Turner, Murray *et al.* (1985).

The prospects for a useful vaccine therefore seem poor. Treatment with antibiotics can be effective, but only if applied early. Managemental measures seem to offer the only hope of control. Such measures should aim at minimizing faecal contamination of the environment and the occurrence of abrasions of the buccal mucosa and skin.

Necrobacillosis in deer and antelope

Numerous outbreaks have been recorded in wild ungulates in North America since the early 1920s. In a review of necrobacillosis in wild mammals Rosen (1981) referred to California, Wyoming, Minnesota, North Carolina, Michigan, Saskatchewan, Ontario and Alaska; and to Rocky Mountain mule deer (*Odocoileus hemionus hemionus*), Columbian black-tailed deer (*Odocoileus hemionus columbianus*), white-tailed deer (*Odocoileus virginianus*), elk (*Cervus canadensis* and *Cervus rooseveltii*), moose (*Alces americana*), caribou (*Rangifer caribou*), pronghorn antelope (*Antilocapra americana*), and bighorn sheep (*Ovis canadensis*).

In the drought years of 1924 and 1925 hundreds of dead mule deer were found in Modoc County, California, in the vicinity of drying springs (Henderson 1925). The deaths, which were associated with the ingestion of large quantities of mud—the only source of water—and with lesions of the mouth, were undoubtedly caused by necrobacillosis (Rosen, Brunetti, Bischoff & Azevedo 1951). Outbreaks in deer reported in California during the next 25 years were often characterized by stomatitis and foot rot. In the 1950 epizootic, which began in June and within four months affected black-tailed deer and mule deer in five counties, foot lesions were rare but stomatitis and rumenitis common (Rosen *et al.* 1951). It appeared that the disease in California was always favoured by dry summers and drought, and that it constituted a serious threat to the continued existence of deer on certain ranges.

Elsewhere in North America, however, outbreaks of necrobacillosis have been seen in winter. Murie (1930) described an epizootic disease of elk at Jackson Hole, Wyoming, in 1928. The snows had driven the animals from the mountains into the valley where they congregated in increasing numbers until supplementary feeding at the Government Elk Refuge began in early January. Of about 7000 elk on and around the feeding grounds 409 (5.8%) died, but the losses in calves (18.4% of 1406 animals) were more severe than in adults. Necrobacillosis was the principal cause of the mortality. Necrotic stomatitis was particularly common, but other manifestations included foot lesions, arthritis, exostoses of the jaw, gastric ulceration, necrotic lesions of the liver and spleen, peritonitis, pneumonia and pleurisy. A further 1046 deaths among the elk of Jackson Hole occurred in the winter of 1942–1943 (Murie, Allred & Brown 1944). Also in the winter, and with a similarly broad range of lesions, necrobacillosis was noted in Saskatchewan in white-tailed deer and pronghorn antelope by Wobeser, Runge & Noble (1975).

Necrobacillosis has also been seen in the game reserves of Africa. A severe outbreak of interdigital necrosis caused by *F. necrophorum* was seen by Dräger (1975) in gemsbok (*Oryx gazella*) in a very localized area ('Deception Pan') of the Central Kalahari Game Reserve, Botswana. About 25% of 450 animals were lame, usually in at least two limbs. The interphalangeal joints were sometimes affected. Three months later 36% of 55 gemsbok observed were still lame. A few (0.6%) of a population of 800 springbok (*Antidorcas marsupialis*) in the same area also became infected. Close observation of 450 gemsbok and 300 springbok in the terrain surrounding Deception Pan revealed no lameness. Severe infection in a gemsbok is illustrated in Fig. 3.

Gainer (1983) drew attention to reports of necrobacillosis in wildebeest

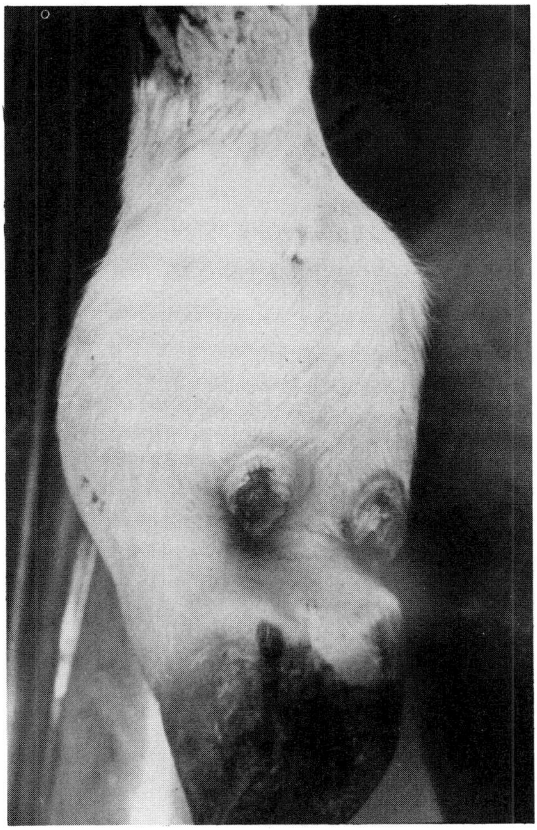

Fig. 3. Foot rot with arthritis in a gemsbok. By courtesy of Professor M. Kanoe.

(*Connochaetes taurinus*) in the Serengeti National Park in 1963 and 1973, and in the Selous Game Reserve, Tanzania, during the wet season in 1970. Gainer himself witnessed an outbreak in the Selous Reserve in 1972 in which lame calves were present in virtually every wildebeest herd during the wet season. *F. necrophorum* was isolated from affected interphalangeal joints. Even mildly affected calves were thought to have little chance of escaping predation, and the disease was considered a significant cause of losses in wildebeest.

Reports from Asia include an account of interdigital necrosis in saiga antelope (*Saiga tatarica*) in the Betpak-Dala desert of Kazakhstan (Petrov, Islamov & Dutov 1979) and numerous references to cases in reindeer (*Rangifer tarandus*), in which serious outbreaks occur in Siberia (Rosen 1981).

Necrobacillosis is sometimes seen in captive wild ungulates. Mettam & Carmichael (1933) reported nine fatal cases of necrotic stomatitis in recently captured young antelope in Uganda. The species affected were the duiker (*Cephalophus* sp.), oribi (*Montana cottoni cottoni*) and reedbuck (*Redunca redunca wardi*). Unhygienic surroundings were thought to have been responsible. Martinaglia (1937) described cases in Johannesburg Zoo in the duiker, steenbok (*Raphiceros campestris*), lechwe (*Onotragus leche*), hog deer (*Cervus porcinus*), and red brocket (*Mazama americana*). The disease took the form of necrotic stomatitis, sometimes with extension to the bone of the jaw; and necrosis of the rumen and adjoining stomachs, usually followed by peritonitis.

Botulism

Botulism is a form of food poisoning affecting mammals, birds and other vertebrates. It is caused by the neurotoxin of *Clostridium botulinum*, a gram-positive spore-forming anaerobe whose natural habitat is soil and mud.

Intoxication and infection

Until a few years ago the disease was generally considered a pure intoxication resulting from the ingestion of toxin formed outside the body on foodstuffs eaten by man or animals. This concept probably still holds good for most cases of botulism. However, ingestion of toxin inevitably implies ingestion of the organism that produced it, and it has always been difficult to rule out under all circumstances an aetiological role for viable *C. botulinum* in the gut. The discovery of 'infant botulism' (Arnon 1980) provided a clear illustration that toxigenesis could occasionally occur in the gut in the presence of a favourable microflora. The importance of such a microflora has been clarified by experiments in very young animals, germ-free animals, and in those treated orally with antibiotics (see Miyazaki & Sakaguchi 1978; Burr, Beery & Sugiyama 1985). In all probability intestinal toxigenesis occurs occasionally in adults as well as in infants (Bartlett 1986).

Toxigenesis in the natural environment

It seems possible that microbiological factors analogous to those concerned in infant botulism play a part in the survival, growth and toxigenic potential of *C. botulinum* in the environment.

This notion is illustrated by the inhibitory properties of Camargue mud for various types of *C. botulinum*, including type C, the cause of botulism in waterfowl (Smith, Moryson & Walmsley 1977; Smith 1978). It has not yet been proved that the inhibitory factors are microbiological, but bacteria (mainly *Bacillus* spp.) with antibiotic activity against *C. botulinum* type C have been found in inhibitory muds (Graham 1978).

Similarly, toxigenesis in rotting carcasses is probably influenced one way or the other by the mixed microflora of carrion. Thus Smith & Turner (1987) found that at 23°C a type C strain of *C. botulinum* produced more toxin in mouse carcasses than in pure broth cultures. This state of affairs was reversed at 37°C. At 30°C similar degrees of toxicity were produced in both systems and at 16°C there was little or no toxigenesis.

Botulism in zoological collections

The important ways in which botulism can occur in zoological collections are illustrated by the following examples.

Spoiled fish are thought to have produced botulism in California sea lions (*Zalophus californianus*) in Kansas City Swope Park Zoo (Wagner & Mann 1978). Type C outbreaks that killed 14 New World monkeys at Bristol Zoo (Smart, Roberts, McCullagh, Lucke & Pearson 1980) and one of six affected lions (*Panthera leo*) at a circus (Greenwood 1985) were caused by spoiled chicken. An outbreak at Chester Zoo in which 39 birds died was associated with a dietary mixture containing minced horse meat (Smith, Brambell, Gaynor, Greed & DeMeurichy 1985).

At the London Zoo type C botulism occurred in birds as a result of feeding commercially bred maggots (Smith, Hime, Keymer, Graham, Olney & Brambell 1975), a practice since discontinued.

At Perth Zoo, Western Australia, three type C outbreaks in the early 1980s caused heavy mortality in the waterfowl collection as well as the deaths of two white-handed gibbons (*Hylobates lar*). These fatal intoxications occurred as a result of the ingestion of rotting carcasses, or the maggots thereon, of wild black ducks (*Anas superciliosa*) that had died in the zoo after ingesting toxin in the natural environment outside (Smith, Brambell *et al.* 1985).

Finally, in recent years type C botulism has occurred on several occasions at Planckendael (Antwerp Zoo) in swans and other water birds owing to toxigenesis in organic debris in slow-moving water (Smith, Brambell *et al.* 1985).

It is unlikely that the few outbreaks of botulism in zoos recorded in the literature faithfully reflect the real incidence of the disease in captive animals.

Botulism in free-living wild animals

Millions of water birds die annually from type C botulism in lakes and marshes throughout the temperate zones of the world, but outbreaks are particularly well known in the western parts of North America (see Smith 1982). The huge scale on which the disease sometimes occurs is illustrated by Rosen's (1971) estimate that between four and five million ducks died from botulism in the western USA in 1952, and his comment that 'This severe mortality undoubtedly exerted a profound influence on the population dynamics governing waterfowl.'

Surprisingly, however, botulism in waterfowl is almost unknown in the tropics. Thus although the disease is recognized in South Africa it has not been recorded in tropical Africa except for a single type D outbreak in Senegal in which 200 pelicans died (Doutre 1979). This may be because the factors that precipitate outbreaks in temperate climates are lacking in the tropics or because, as seems more likely, outbreaks in the tropics remain unreported or undiagnosed. It is worth recalling that in the UK botulism in water birds, which is now well known in warm summers, was unrecognized less than 20 years ago. *C. botulinum* type C appears to be widespread in tropical inland waters as indicated by surveys made, from the Institute of Zoology, in Mauritius, the Seychelles, Nigeria and tropical Botswana. Perhaps surprisingly, however, a thorough examination of Lake Nakuru, Kenya, in 1984 failed to reveal the organism (G. R. Smith, unpublished).

C. botulinum antitoxin has been found in the serum of carrion-eating birds and mammals (Holdeman 1970; Ohishi, Sakaguchi, Riemann, Behymer & Hurvell 1979). There are no reports of botulism in free-living wild mammals, but it would seem certain that cases must occur, probably on a large scale. As in domesticated mammals, the main source of botulinum toxin for wild mammals is probably carrion. The laboratory experiments already mentioned (Smith & Turner 1987), if applicable to the field, suggest that toxigenesis in carrion is favoured less by temperatures encountered in the more extreme tropical climates than by those of a warm English summer.

Concluding remarks

Both diseases discussed cause significant losses in wildlife but, bearing in mind waterfowl mortality, there is little doubt that of all the microbiological diseases of free-living animals botulism takes the heaviest annual toll.

Strangely, however, almost nothing is known of wildlife botulism in tropical zones. This obvious gap in knowledge invites serious study. Such study, as well as being important in its own right, would almost certainly bring investigators into contact with other disease problems. At present research on tropical wildlife diseases seems undervalued, and opportunities thrown up by chance should be exploited to the full.

Acknowledgement

The support of the Wellcome Trust for research on necrobacillosis and botulism at the Institute of Zoology is gratefully acknowledged.

References

Arnon, S.S. (1980). Infant botulism. *A. Rev. Med.* **31:** 541–60.

Ashton, D.G. (1982). Veterinary clinical report. Whipsnade Park. (In The Zoological Society of London Scientific Report 1979–81). *J. Zool., Lond.* **197:** 82–100.

Bartlett, J. C. (1986). Infant botulism in adults. *New Engl. J. Med.* **315:** 254–5.

Beerens, H., Fievez, L. & Wattre, P. (1971). Observations concernant 7 souches appartenant aux espèces *Sphaerophorus necrophorus, Sphaerophorus funduliformis, Sphaerophorus pseudonecrophorus. Annls Inst. Pasteur Lille* **121:** 37–41.

Beveridge, W.I.B. (1934). A study of twelve strains of *Bacillus necrophorus*, with observations on the oxygen intolerance of the organism. *J. Path. Bact.* **38:** 467–91.

Burr, D.H., Beery, J.T. & Sugiyama, H. (1985). Importance of cecum in *Clostridium botulinum* colonization of mice and relationship of organism to large bowel. *Curr. Microbiol.* **12:** 277–82.

Burton, J.D. (1981). *Studies into aspects of lumpy jaw in macropods.* PhD Thesis: University of Melbourne.

Doutre, M.P. (1979). Botulisme de type D chez des pélicans (*Pelecanus rufescens*) lié à un changement du milieu naturel (Petite Côte, Senegal). *Malimbus* **1:** 110–13.

Dräger, N. (1975). A severe outbreak of interdigital necrobacillosis in gemsbok (*Oryx gazella*) in the northern Kalahari (Botswana). *Trop. Anim. Hlth Prod.* **7:** 200.

Evans, J.W. & Berg, J.N. (1985). Development of enzyme-linked immunosorbent assays for the detection of *Fusobacterium necrophorum* antibody in animal sera. *Am. J. vet. Res.* **46:** 132–5.

Finnie, E.P. (1976). Necrobacillosis in kangaroos. In *Wildlife diseases*: 511–8. (Ed. Page, L.A.). (Proc. 3rd int. Wildlife Conf., Munich 1975). Plenum Press, New York.

Fox, H. (1923). *Disease in captive wild mammals and birds. Incidence, description, comparison.* J. B. Lippincott, Philadelphia, London and Chicago.

Gainer, R.S. (1983). Necrobacillosis in wildebeest calves. *J. Wildl. Dis.* **19:** 155–6.

Graham, J.M. (1978). Inhibition of *Clostridium botulinum* type C by bacteria isolated from mud. *J. appl. Bact.* **45:** 205–11.

Greenwood, A.G. (1985). Diagnosis and treatment of botulism in lions. *Vet. Rec.* **117:** 58–60.

Hawkey, C.M., Ashton, D.G., Hart, M.G., Cinderey, R.N. & Jones, D.M. (1982). Haematological findings in red-necked wallabies (*Protemnodon rufogrisea*) with necrobacillosis. *Res. vet. Sci.* **33:** 290–4.

Henderson, C.A. (1925). Modoc deer disease. *Calif. Fish Game* **11:** 27–30.

Hofstad, T. (1985). *Fusobacterium necrophorum*—pathogenic organism? *J. med. Microbiol.* **20:** vii.

Holdeman, L.V. (1970). The ecology and natural history of *Clostridium botulinum. J. Wildl. Dis.* **6:** 205–10.

Jensen, R., Deane, H.M., Cooper, L.J., Miller, V.A. & Graham, W.R. (1954). The rumenitis-liver abscess complex in beef cattle. *Am. J. vet. Res.* **15:** 202–16.

Keane, C., Taylor, M.R.H., Wilson, P., Smith, L., Cunningham, B., Devine, P. &

English, L.F. (1977). *Bacteroides ruminicola* as a possible cause of 'lumpy jaw' in Bennett's wallabies. *Vet. Microbiol.* **2:** 179–83.

Martinaglia, G. (1937). Some considerations regarding the health of wild animals in captivity. *S. Afr. J. Sci.* **33:** 833–44.

Mettam, R.W.M. & Carmichael, J. (1933). Necrobacillosis in recently captured antelope in Uganda. *J. comp. Path. Ther.* **46:** 16–24.

Miyazaki, S. & Sakaguchi, G. (1978). Experimental botulism in chickens: the cecum as the site of production and absorption of botulinum toxin. *Jap. J. Med. Sci. Biol.* **31:** 1–15.

Murie, O.J. (1930). An epizootic disease of elk. *J. Mammal.* **11:** 214–22.

Murie, O.J., Allred, W.J. & Brown, R.C. (1944). Disease kills feedground elk. *Wyo. Wildl.* **9:** 1–8, 27.

Ohishi, I., Sakaguchi, G., Riemann, H., Behymer, D. & Hurvell, B. (1979). Antibodies to *Clostridium botulinum* toxins in free-living birds and mammals. *J. Wildl. Dis.* **15:** 3–9.

Oliphant, J.C., Parsons, R. & Smith, G.R. (1984). Aetiological agents of necrobacillosis in captive wallabies. *Res. vet. Sci.* **36:** 382–4.

Petrov, V.F., Islamov, R.Z. & Dutov, M.M. (1979). Necrobacillosis in saiga antelopes. *Veterinariya* **1979**(3): 48.

Roberts, D.S. (1969). Synergic mechanisms in certain mixed infections. *J. infect. Dis.* **120:** 720–4.

Rosen, M.N. (1971). Botulism. In *Infectious and parasitic diseases of wild birds*: 100–17. (Eds Davis, J.W., Anderson, R.C., Karstad, L. & Trainer, D.O.). Iowa State University Press, Ames.

Rosen, M.N. (1981). Necrobacillosis. In *Infectious diseases of wild mammals*: 332–8. (Eds Davis, J.W., Karstad, L.H. & Trainer, D.O.). Iowa State University Press, Ames.

Rosen, M.N., Brunetti, O.A., Bischoff, A.I. & Azevedo, J.A. (1951). An epizootic of foot rot in California deer. *Trans. N. Am. Wildl. Conf.* **16:** 164–79.

Samuel, J.L. (1983). Jaw disease in macropod marsupials: bacterial flora isolated from lesions and from the mouths of affected animals. *Vet. Microbiol.* **8:** 373–87.

Scott, H.H. (1925). A streptothrix disease of wallabies. *Proc. zool. Soc. Lond.* **1925:** 799–814.

Simon, P.C. & Stovell, P.L. (1969). Diseases of animals associated with *Sphaerophorus necrophorus*: characteristics of the organism. *Vet. Bull. Weybridge* **39:** 311–15.

Smart, J.L., Roberts, T.A., McCullagh, K.G., Lucke, V.M. & Pearson, H. (1980). An outbreak of type C botulism in captive monkeys. *Vet. Rec.* **107:** 445–6.

Smith, G.R. (1978). Botulism, waterfowl and mud. *Br. vet. J.* **134:** 407–11.

Smith, G.R. (1982). Botulism in waterfowl. *Symp. zool. Soc. Lond.* No. 50: 97–118.

Smith, G.R., Brambell, M.R., Gaynor, W.T., Greed, G.R. & DeMeurichy, W. (1985). Botulism in zoological collections. *Vet. Rec.* **117:** 534.

Smith, G.R., Hime, J.M., Keymer, I.F., Graham, J.M., Olney, P.J.S. & Brambell, M.R. (1975). Botulism in captive birds fed commercially bred maggots. *Vet. Rec.* **97:** 204–5.

Smith, G.R., Moryson, C.J. & Walmsley, J.G. (1977). The low prevalence of

Clostridium botulinum in the lakes, marshes and waterways of the Camargue. *J. Hyg., Camb.* **78:** 33–7.

Smith, G.R., Oliphant, J.C. & Parsons, R. (1984). The pathogenic properties of *Fusobacterium* and *Bacteroides* species from wallabies and other sources. *J. Hyg., Camb.* **92:** 165–75.

Smith, G.R. & Turner, A. (1986). The adverse effect of dilution on the infectivity of *Fusobacterium necrophorum* culture. *J. Hyg., Camb.* **96:** 199–203.

Smith, G.R. & Turner, A. (1987). Factors affecting the toxicity of rotting carcasses containing *Clostridium botulinum* type C. *Epidem. Inf.* **98:** 345–51.

Smith, G.R., Turner, A. & Cinderey, R. (1986). Susceptibility of wallabies to *Fusobacterium necrophorum*. *Vet. Rec.* **118:** 691–3.

Smith, G.R., Turner, A., Murray, L.G. & Oliphant, J.C. (1985). The weak immunogenicity of *Fusobacterium necrophorum*. *J. Hyg., Camb.* **95:** 59–68.

Tomlinson, A.R. & Gooding, C.G. (1954). A kangaroo disease: investigations into 'lumpy jaw' on the Murchison, 1954. *J. Agric. West. Aust.* **3:** 715–18.

Turner, A., Bidwell, D. & Smith, G.R. (1986). The serological response of mice to *Fusobacterium necrophorum*. *Res. vet. Sci.* **41:** 412–13.

Tyndale-Biscoe, H. (1973). *Life of marsupials*. Edward Arnold, London.

Wagner, J.E. & Mann, P.C. (1978). Botulism in California sea lions (*Zalophus californianus*)—a case report. *J. Zoo Anim. Med.* **9:** 142–5.

Warner, J.F., Fales, W.H. & Teresa, G.W. (1974). Passive haemagglutination test for determining the immune response of rabbits to *Sphaerophorus necrophorus* of bovine hepatic origin. *Am. J. vet. Res.* **35:** 551–4.

Watts, P.S. & McLean, S.J. (1956). Bacteroides infection in kangaroos. *J. comp. Path. Ther.* **66:** 159–62.

Wobeser, G., Runge, W. & Noble, D. (1975). Necrobacillosis in deer and pronghorn antelope in Saskatchewan. *Can. vet. J.* **16:** 3–9.

Symp. zool. Soc. Lond. (1988) No. 60 175–199

Viruses transmissible between wild and domestic animals

Walter PLOWRIGHT

Whitehill Lodge
Goring-on-Thames
Reading
Berks RG8 0LL

Synopsis

The importance of wildlife diseases has received grudging, if any, recognition from most biologists and conservationists. This is understandable in that research has been predominantly concerned with agents communicable from wildlife to man or domestic animals. However, little wildlife disease research would have been undertaken without the interest and finance generated by concern for the health of man and his domestic animals.

This paper considers, in turn, a few examples of virus infections of wildlife which can be regarded as of biological interest because:

1. They contribute to the regulation of populations (rinderpest and canine distemper).
2. They are reservoirs of disease in man and domestic animals, explaining events in the past and providing warnings for the future (infections due to orthopoxviruses, avian paramyxoviruses and influenza viruses).
3. They have comparative epidemiological interest in showing how virus infections behaved in their original hosts prior to their establishment in domesticated secondary hosts (foot-and-mouth disease and calicivirus infection).
4. They have implications for animal welfare and captive breeding programmes in zoos as well as for conservation in the wild (infections due to the herpesviruses of malignant catarrhal fever).

Many of the examples could obviously be placed in more than one category. The lack of information about the effects of these infections on free-living populations is a common deficiency which will be difficult and expensive to eliminate; it also points to the need for better collaboration between biologists and pathologists.

Introduction

The subject of this paper is probably, at first sight, one of restricted scope and of relatively little importance in developed economies. My own interest arose at the end of the 1950s in East Africa, when I began to investigate the role of wildlife or 'game animals', as they were then called, in the epizootiology of malignant catarrhal fever and rinderpest of cattle. This exercise

ZOOLOGICAL SYMPOSIUM No. 60
ISBN 0-19-854007-8

began in the Kajiado District of Kenya, next to the Nairobi National Park, and soon extended to the Ngorongoro Conservation Area and Serengeti National Park in Tanzania. In such surroundings it was easy to appreciate the relevance and attraction of wildlife studies. Furthermore, the basic tools for such investigations—cell culture techniques for virus isolation and serological surveys—were just becoming available. These tools made objective and quantitative studies feasible, where previously they had been virtually impossible.

If blatant anthropomorphic prejudice was the dominant reason for wildlife disease research in Africa 20–30 years ago, it was hardly different elsewhere. For example, rabies in foxes or other wildlife was, as now, primarily of interest because it infected man (very rarely) or his domestic animals (much more frequently). What I would like to illustrate in this paper is that, in respect to virus infection, wild animals are much more than passive reservoirs of unpleasant and economically serious diseases of man and his livestock. I lay no claim to originality in this approach or to virtue in having 'seen the light'. A critique and philosophy of research in wildlife diseases generally was clearly formulated by Milton Friend (1976), at the opening of the Third International Wildlife Disease Conference in 1975, as follows: '. . . resistance or apathy towards controlling wildlife disease arises because (1) seldom is this action initiated primarily for the direct benefit of wildlife; (2) the action may be at the direct expense of wildlife; (3) there has been little to demonstrate that wildlife diseases can be controlled in free-living populations; (4) the role of disease in the population dynamics of wildlife species has not been adequately documented . . . '

For the manager of wildlife resources, from whose viewpoint Friend was attempting to summarize the position, it could well be that the order of importance of his four deductions should now be inverted; they are all, nevertheless, still highly relevant.

The title of my paper is somewhat misleading—it is not the intention to list the viruses transmissible from wild to domestic animals or in the opposite direction. A close examination will usually reveal that there are free-living hosts for viruses that affect domestic animals; sometimes we shall see that the former were in all probability the original hosts as, for example, with the caliciviruses of vesicular exanthema, the picornaviruses of the SAT types of foot-and-mouth disease or the paramyxo- and myxoviruses causing Newcastle disease and fowl plague. In some diseases, such as distemper and rinderpest in Africa (Neitz 1967) and louping ill in grouse in Scotland (Reid 1984), the evidence strongly suggests that domestic animals served to introduce viruses to hitherto uninfected wildlife. We do not have an archaeological or geological record to trace the evolution of viruses, but the present flood of information on viral genomes has given rise to waves of speculative interest. It is not wildly improbable that many of the major groups of animal (and human) viruses emerged from free-living species, with domestication and pre-civilization; some are still emerging, as we shall see with calici-viruses, not to mention influenzaviruses and HIV1.

I propose to discuss some examples of viruses that have several broad attributes of interest to both biologist and pathologist:

1. Those producing diseases that act as regulatory factors, especially on populations of rare species.
2. Those having reservoirs in wild hosts and causing important diseases of domestic animals (or man).
3. Viruses having comparative scientific interest.
4. Viruses important for animal welfare, whether in the wild or zoological collections.

Viruses contributing to population control

Myxomatosis

Certain viruses have a profound effect on animal populations. This has sometimes been exploited—as in myxoma in rabbits—for the control of animal pests. The myxoma virus in its original American host (*Sylvilagus* spp.) generally caused localized lesions only (see Yuill 1981; Fenner 1983; Ross 1982) and it is not known to have had any effect on population; it was only when virulent virus and suitable vectors were deliberately brought together by man with highly susceptible *Oryctolagus* populations that major epidemics, with enormous mortality and secondary ecological changes (R.M. Anderson & May 1986), were produced in Australia and Europe.

Rinderpest in the Serengeti region

Another example is rinderpest, which invaded Africa from the north-east from about 1889, reaching the Cape by the end of 1897; some highly susceptible species, such as buffaloes, were reduced to one tenth or less of their original numbers. Others such as greater kudu, eland and bongo were so reduced in numbers that some populations never recovered or were eliminated (G. R. Scott 1981; Plowright 1982). The disease was eradicated from the south of the continent but persisted north of the equator and became enzootic from the late 1920s or early 1930s in wildlife of the 'Serengeti region' of northern Tanzania and southern Kenya. This persistent focus caused periodic mortality during the early 1930s in species such as buffalo (*Syncerus caffer*) and blue wildebeest (*Connochaetes taurinus*), but it was not regarded by game wardens as important. It was, however, regarded as a source of infection for cattle. It caused deaths primarily in yearlings and two year old animals in the Serengeti area, indicating that older animals had been infected and were immune (see Plowright & McCulloch 1967). The annual or biennial mortality was reported in wildebeest up to 1960, both in the Serengeti National Park and in the adjacent Mara area of Kenya, where the disease was observed by Talbot & Talbot (1961) and investigated by Plowright and McCulloch serologically from 1960 onwards. Similar investigations in the Ngorongoro and Serengeti districts began in 1959 and the last clinical cases in wildlife attributable to rinderpest occurred there in 1961 or 1962, as also in the Mara district. The last year during which widespread serological conversion took place was 1963. No rinderpest was reported in cattle after 1966, when a small focus with minimal spillover into wildlife near the Serengeti Park was confirmed (Taylor & Watson 1967). From then until 1971, serological surveys failed to established any rinderpest foci in the wildlife or cattle of southern Kenya or Tanzania and the disease was considered to have been eradicated.

Based primarily on this evidence the first wave of an eruption of the wildebeest and buffalo populations of the Serengeti region, which took place between 1961 and 1971, was attributed to the disappearance of rinderpest. The census figures are impressive—a doubling of the wildebeest population took place between 1961 and 1967 (250,000 to 500,000) and by 1977 it had increased seven-fold. A delayed but comparable increase occurred in buffalo (30,000 to 50,000) during the period 1961–67, but the numbers of zebra barely changed (Sinclair 1977; Sinclair & Norton-Griffiths 1979). The contention of Spinage (1962), repeated by R. M. Anderson & May (1986), that the effect of rinderpest on prey animals, such as wildebeest and buffalo, in East Africa was severe enough to cause a substantial reduction in predators such as lions, leopards and hunting dogs, seems to be at variance with other observations. For example, in Etosha (Namibia) heavy mortality due to anthrax in wildebeest and zebra was associated with an increase in large predators (lion and hyena), who were provided with readily available food (Berry 1981).

The sequel to this story has now partly emerged. Rinderpest reappeared in buffaloes of the Lobo area of the Serengeti National Park in March and April 1982, and several hundred animals died; a further outbreak in buffaloes occurred in June–August 1982 in the Ngorongoro Crater, and about 1000 of 3000 animals probably died. It was later confirmed that virtually all buffaloes born before these incidents developed rinderpest neutralizing antibody. Seroconversion, without clinical signs, also occurred at about the same time in the great majority of buffaloes in the Mara area of Kenya, but infection apparently died out, first on the western side and, by late 1984, on the east. A survey in June 1986 showed that nearly all Serengeti buffaloes in widely separated localities continued to be infected until 1984, frequently without reports of clinical disease. In these later incidents wildebeest did not become infected at all (Rossiter 1986; Rossiter, Jessett *et al.* 1983; Rossiter, Taylor *et al.* 1987).

It was established in the 1960s that rinderpest, often regarded as a highly lethal disease, frequently caused subclinical infections in wildebeest and other species such as eland or Thomson's gazelle prior to 1963 (Plowright & McCulloch 1967). The strains of virus current up to that time were of reduced virulence for cattle, and many buffaloes, often regarded as highly sensitive, obviously recovered (see Plowright 1982). When rinderpest reappeared in 1982, perhaps earlier in 1980–81, the virus was at first apparently of high virulence for buffaloes in the Serengeti area, but later both there and in neighbouring areas of Kenya the disease may have become clinically inapparent. It has been postulated that a very mild strain or even a rinderpest-like virus is involved and it certainly has a reduced natural host range, but still infects cattle (Rossiter, Jessett *et al.* 1983; Rossiter, Taylor *et al.* 1987).

This failure to observe signs of disease need not surprise us unduly. Approximately 13,000 and 16,000 juvenile buffaloes died in the Serengeti in the absence of rinderpest in the years 1970/71 and 1971/72 respectively at a time when the total population was about 61,000 (Sinclair 1977). Yet very few of these were found moribund, in spite of close observation, presumably because their demise was hastened by predators. It required a retrospective analysis of wildlife population trends, covering a period when rinderpest was shown by serological surveys to have been continuous and widespread and then suddenly disappeared, to indicate what a profound limiting effect the disease had exerted on wildebeest and buffalo numbers for over 30 years. It will be very interesting to follow the possible effects of the present epizootics on buffalo populations.

The epidemiological investigations required cannot be simple. The 'susceptible' populations comprise many species diverse in their genetic susceptibility to the virus, and in their distribution, behaviour, and seasonal reproductive activity. The virus strains may vary in their ability to spread and to produce disease in the possible range of hosts; both properties can be influenced by a multitude of intercurrent viral, protozoal or parasitic infections (see, for example, theileriasis and bovine herpesvirus 2 infection: Schiemann, Plowright & Jessett 1971).

Canine distemper in the black-footed ferret

A recent example of lethal disease, caused by a virus closely related to that of rinderpest and threatening extinction of a very rare species, is that of canine distemper virus in the black-footed ferret (*Mustela nigripes*). This species had been reduced to vanishing point, probably following a decline in the prairie dog (*Cynomys gunnisoni*) population, on which it depends for prey; it is thought, incidentally, that this in turn was due, at least partly, to sylvatic plague (see Barnes 1982). Then a colony of black-footed ferrets was found in an area of 80,000 acres in Meeteetse, Wyoming, in September 1981; it prospered, so that by 1984 it comprised 128 members. By August 1985, however, the numbers had dropped alarmingly (to 58) and on 13 September were down to 29, with two taken into captivity. Sylvatic plague was considered as a possible cause. On 22 October six ferrets were taken into captivity and one subsequently died from confirmed canine distemper. Further sick and healthy ferrets were captured but died of distemper, as did the majority of those left in the wild.

An emergency trapping operation then yielded six ferrets. These were placed in isolation and given an inactivated vaccine with adjuvant, live attenuated vaccine having previously been found to be lethal (Carpenter, Appel, Erickson & Novilla 1976). There may be a few, perhaps three, ferrets left in the wild, but the future of the species now depends on a captive breeding programme—a measure that was apparently vetoed in 1984 by the Wyoming Game and Fish Department but is now funded to a figure of $389,000 (Thorne 1986; see also May 1986). The discovery of distemper in captured animals was a fortunate and 'in a sense, fortuitous' discovery, but it should be a warning of the necessity for 'special concern about the possibility of disease in the case of a species that is rare and yet whose population is locally undergoing steady growth' (May 1986). After all, ferrets have been used as highly susceptible experimental animals for canine distemper research since the classical studies of Dunkin & Laidlaw (1926), and in Wyoming many species such as raccoons and coyotes could have brought distemper virus to the ferrets.

The ferret is not the only species that could be threatened by the ravages of canine distemper virus. In the Serengeti region the Cape hunting or wild dog (*Lycaon pictus*) has suffered a rapid decline since the late 1960s (Frame, Malcolm, Frame & Lawick 1979), and distemper was reported to be associated with a 'crash' in the population of silver-backed jackals (Moehlman 1983). As bat-eared foxes (*Otocyon megalotis*), mongooses and hyenas are also known to be susceptible, it should be rewarding to study the epidemiology and effects of distemper virus on all the species at risk in this area. Furthermore, it is probable that some of them are susceptible to a range of lethal virus diseases, such as rabies, canine hepatitis and canine parvovirus infection, which could be introduced to them by domestic dogs or other wildlife in or near the National Parks. Viruses are not by any means the only possible pathogens; deaths attributed to *Rickettsia canis* in Cape hunting dogs in the Kruger National Park in the 1930s were thought to be related to a conspicuous decline in the population (Du Toit 1947; Neitz 1967). It should no longer be acceptable to wait until a major epidemic in a restricted species becomes evident from reports of carcases or from population counts before disease investigations are undertaken.

Viruses whose natural reservoirs are probably in wild hosts

Some poxvirus infections of animals

The recent history of 'cowpox' is a fascinating one, especially as regards 'unusual' victims and the probable reservoirs in wild species (see Baxby 1977a). Orthopoxviruses with similarities to smallpox and cowpox or vaccinia were isolated from the kidney tissue of two monkeys and two rodents (*Mastomys coucha* and *Heliosciurus rufobrachium*) in Zaire during the period 1971–75. The last case of smallpox occurred in Central Africa in 1970 and it therefore appeared that these species might be reservoir hosts of the viruses, although convincing serological evidence was not obtained. Similar orthopoxviruses (whitepox variants) were isolated in 1964 in Holland from the kidneys of two apparently normal *Cynomolgus* monkeys and cowpox was isolated from an outbreak in okapi (*Okapia johnstonii*) at the Rotterdam zoo, in which one of five affected animals died. Between 1961 and 1977 there were six outbreaks of a poxvirus disease, sometimes fatal, in elephants in zoos in East and West Germany; the virus, which was vaccinia-like, was called 'elephant pox', but it was soon recognized to be similar to, though separate from, cowpox. The origin of these outbreaks was not established, but the disease is now controlled in elephants by vaccination (Baxby & Ghaboosi 1977).

In 1973–74 two outbreaks of an acute, often lethal, disease occurred in large Felidae at the Moscow zoo, affecting lion, black panther, cheetah, puma, jaguar and ocelot, as well as giant anteaters (*Edentata*). Again, viruses closely related to cowpox but not identical with it were isolated; the source of infection was a farm that produced live white rats as food for the zoo and on which there had been two serious epizootics of an identical infection in the three preceding months (Marennikova, Maltseva, Korneeva & Garanina 1977). Identical viruses were isolated in 1974 from big gerbils (*Rhombomys opimus*) in Turkmenia and specific antibody capable of inhibiting both haemagglutination and infectivity was found in this species, as well as in yellow susliks (*Citellus fulvus*), which were also highly susceptible experimentally (Marennikova, Ladnyj, Ogorodnikova, Shelukhina & Maltseva 1978).

In 1977, cowpox virus was associated with very severe disease in cheetahs at Whipsnade; three of nine were infected and two died (Baxby, Ashton, Jones & Thomsett 1982). A second outbreak occurred in another British zoo in 1978, in which two infected cheetahs died; at the same time a domestic kitten died of confirmed cowpox infection (Baxby, Ashton, Jones, Thomsett & Denham 1979). Meanwhile, Thomsett, Baxby & Denham (1977) had described cowpox infection in the domestic cat and Baxby (1977b) had pointed out that cowpox infection in man was sometimes severe and that it was by no means always associated with contact with diseased cattle. It was 'not known, in fact, whether cows had ever been a reservoir of cowpox virus' even in Jenner's day; between 1964 and 1975 there were only 13 confirmed cases of bovine cowpox in Britain and, furthermore, the causative viruses were not all of the same biotype. Hence, there was a problem in explaining the maintenance of virus and it was more than a possibility that other reservoir hosts played a role.

More cases of 'cowpox' infection in domestic cats have continued to be reported in Britain (Martland, Fowler, Poulton & Baxby 1983; Martland, Poulton & Done 1985; Hoare, Gruffydd-Jones, Bennett, Gaskell & Baxby 1984; Martin, Scott, Lauder & Nash 1984). One group of workers observed 55 cases (Bennett, Gaskell, Gaskell, Baxby & Gruffydd-Jones 1986) and stressed that the primary lesions were usually of such location and appearance as to suggest infection by bite wounds; cat-to-cat spread is very unusual and evidence strongly suggests that domestic cats, like the larger Felidae, are indicator, not maintenance hosts. The felid isolates are indistinguishable from classical cowpox (Bennett, Baxby, Gaskell, Gaskell & Kelly 1985) but Martin *et al.* (1984) found that one was of very low pathogenicity in the skin of cattle. Natural transmission of feline virus to man has been reported in the Netherlands (Willemse & Egberink 1985) and in Britain (Pether, Trevains, Harrison, Baxby, Bennett & Gibb 1986) but appears to be rare.

The predominance of cases in cats in rural areas in autumn could indicate infection from small mammals, which are most numerous and active at this season (Bennett, Gaskell *et al.* 1986), but the direct evidence for this (Kaplan, Healing, Evans, Healing & Prior 1980) is limited to the finding of orthopoxvirus antibody in seven short-tailed voles (*Microtus agrestis*) out of 20 examined, and in two out of 28 woodmice (*Apodemus sylvaticus*).

The disease in large Felidae and elephants is obviously a matter of concern to zoo managers and the reported failure of vaccination in cheetahs (Baxby *et al.* 1979, 1982) lends further emphasis to the need for epidemiological enquiry. The cheetah is a threatened species, for which evidence of genetic monomorphism is already available. When a group of Felidae in an Oregon sanctuary was exposed to the coronavirus of feline infectious peritonitis, all of 42 cheetah became infected within six months, 38 of them clinically and 18 fatally; lions were apparently resistant (O'Brien *et al.* 1985). Cowpox virus also needs to be added to the list of agents that might threaten cheetah breeding programmes.

Before leaving the poxviruses it is necessary to mention briefly one that has yet received little microbiological investigation, probably because it is not known to be transferable to man or domestic species; it nevertheless demands attention as being associated with the spectacular decline of the red squirrel in England. A poxvirus disease of this species has been recognized in Britain (and Finland) on clinico-pathological grounds since the early 1900s (Vizoso 1968). It has been attributed to an agent with the properties of a myxovirus and later to a virus resembling, but in some respects differing from, a parapoxvirus—i.e. a virus similar to that of orf in sheep (A. C. Scott, Keymer & Labram 1981). A pathologically similar disease in sealions and harbour seals, caused by a virus of the paravaccinia (parapox) group has been reported in American waters (Karstad 1981). The causal agent has now been isolated in cell cultures from a red squirrel (Sands, Scott & Harkness 1984) but is still poorly characterized and its epidemiology remains to be elucidated. Infection is not apparently dependent on introduction by grey squirrels, which are not clinically affected, and there is evidence that the disease can persist in 'recovered' populations after a severe outbreak. Reynolds (1985) considered that the disease required further intensive investigation but concluded that 'both disease and island effects have clearly been important features of species replacement', viz. of the red by the grey squirrel.

Avian reservoirs of viruses in wildlife

One of the most remarkable epidemiological stories to unfold in the last decade is that of the role of wild birds, particularly migratory waterfowl and psittacines, as vast reservoirs for the genetic material of myxoviruses and paramyxoviruses. These include not only serious pathogens for man (influenza) but also for domestic and wild animals (fowl plague and Newcastle disease of poultry; seal influenza).

Avian paramyxoviruses

The early history of Newcastle disease (avian paramyxovirus 1, A-PMV 1), was one of isolated but severe epizootics near seaports, as in the Dutch East Indies, Korea and north-west England. It was followed by the establishment of milder forms in the 1940s in Europe, Africa and South America. A second epizootic, in the 1960s and early 1970s, was characterized by highly virulent (viscerotropic, velogenic; or 'VV') strains, apparently spread by the international trade in psittacine birds from South East Asia and South America. Game birds, particularly pheasants and pigeons, also played a part in local dissemination of these viruses, as well as suffering severely in some instances (Hanson 1976).

In addition, waterfowl on the eastern, central and western flyways in the United States, migrating between the arctic and Caribbean regions, were found to be inapparently infected with mild (lentogenic) strains of virus. These agents were often detected in cloacal swabs, indicating that faecal excretion was the common route of dissemination. The extensive trade in tropical cage birds, such as canaries, budgerigars, finches and pittas, is also a potential means of introducing virulent stains of virus (Hanson 1976). In recent times in the USA strains of VV-NDV have been introduced annually by shipments, frequently illegal, of yellow-naped Amazon parrots (United States Department of Argiculture Dec. 1983, Sept. 1984, June 1985), especially into California and other southern states of the USA; transmission to domestic poultry has nevertheless occurred rarely. During 1980 and 1981, about half a million exotic birds were imported into Great Britain and of these over 54,000 died in quarantine; from them 13 isolates of NDV were obtained of which eight were highly virulent but only two came from parrots (Alexander, Allan, Parsons & Collins 1982). It has been suggested, on the basis of examination of birds in quarantine, that highly virulent NDV may be endemic in South American psittacines (Senne, Pearson, Miller & Gustafson 1983), but no direct epidemiological studies have been reported.

During the period 1981–83 a neurotropic form of Newcastle disease in racing pigeons spread through Europe, probably from the south-east (Iraq and Egypt); it reached Britain in June 1983 and became widespread in both racing and feral stocks, as well as in chicken flocks (Alexander, Lister & Wilson 1986). It was later found (Lister, Alexander & Hogg 1986) that the majority of outbreaks in domestic chickens were associated with foodstuffs contaminated by sick feral pigeons in storehouses at the Liverpool docks or in vehicles and other items used in food transport and processing. The strains of virus were identified serologically, especially by means of monoclonal antibodies (Alexander, Russell & Collins 1984; Alexander, Wilson, Russell, Lister & Parsons 1985). They were found to be distinct from the more classical isolates of A/PMV 1 but differed in only one epitope from four viruses isolated from wild ducks in Europe during 1976–79—a finding possibly of phylogenic significance (Alexander, Russell *et al.* 1984). They also multiplied in, and were excreted primarily from, the gut. Up to the time of this epizootic it had been assumed that Newcastle disease in Columbidae resulted from spread from domestic fowl. After the disease became established in racing pigeons, infection of feral birds was regarded in Europe as representing secondary spread.

Newcastle disease virus (A-PMV-1) is only the first of a group of avian paramyxoviruses, which now comprises at least nine members (Alexander 1982a; Alexander, Hinshaw, Collins & Yamane 1985). Many have been isolated from free-living birds. Some have wide geographical and host ranges, but others are more restricted. Alexander (1982a) concluded that 'several of the avian paramyxoviruses groups, especially PMV–2 viruses, may be naturally endemic in the populations of small wild birds or waterfowl in many countries' and 'only the tip of the iceberg of avian paramyxovirus ecology has been seen'.

Avian myxoviruses

Fowl plague occurred throughout the world in the early part of this century and was often treated as a severe disease requiring slaughter policies for its control. Nevertheless the responsible agent was not recognized as an influenza A virus (orthomyxovirus) until 1955. Subsequently other viruses, similar but sometimes of much reduced virulence, were often isolated from domestic poultry and, occasionally, from wild birds. Influenza A viruses all have a similar nucleoprotein core but differ in the external projections or spikes on the virions, which can be one of 13 distinct haemagglutinins, or one of nine different neuraminidases. The various combinations of these two components—not the host species of origin—are the basis for classification of the influenza viruses. In the last decade or so the frequency of isolation of influenza viruses from turkeys has increased, especially in North America, but severe losses have also been reported from several other countries. Close virological investigation has revealed that flocks of domestic ducks may also harbour one or more influenza A viruses, with or without associated respiratory signs; for example, 290 isolates were obtained from ducks in Hong Kong between 1975 and 1978, representing 37 different antigenic combinations (*vide supra*).

Epidemiological studies of avian influenza virus infections in domestic birds usually indicate an independent origin of outbreaks, i.e. secondary spread by human or other agencies is uncommon and there is very little chance of long-distance spread, for example by air currents. In addition, the strains of virus that occur in isolated outbreaks are often distinct antigenically and in host range. Arising from these observations is the widespread belief that virus introductions are usually attributable to wild birds. This would explain why the regions of maximum incidence, such as Michigan (USA), Israel and Norfolk (Great Britain) often lie on migration routes for wildfowl. Turkeys and ducks are frequently kept in open or partly enclosed pens, in which contact with wild birds is possible; in contrast, chickens are usually strictly confined and, hence, less frequently infected. In Canada, at least, the adoption of measures to prevent contact between wild birds and turkeys has led to a marked decline in the incidence of influenza.

Influenza virus infection has been recorded in 17 different families of wild birds, but the most frequent hosts are Anatidae, especially migratory species (Alexander 1982b). The first reported isolation was of the H5N3 antigenic type, from severely affected common terns (*Sterna hirundo*) in South Africa in 1961. Dead birds were found along a 1000-mile long coastline; 1300 were counted in four small localities (Becker 1966). Subsequently, however, seabirds have been less prolific than waterfowl as a source of influenza viruses. The difficulties in obtaining adequate epidemiological data and information about pathology in wild birds were, however, emphasized by Slemons & Easterday (1976).

Rates of infection in wild ducks in North America vary from 18–60% for juvenile birds and 4–27% for adults (Hinshaw, Webster & Turner 1980). The strains from waterfowl have little or no virulence for their hosts of origin but are of variable virulence for domestic poultry. Virus is excreted from the intestine rather than by the usual respiratory route. The quantity shed in faeces into lake water is so great, and the stability of the virus so high (over 30 days at 0°C, 4 days at 22°C), that direct isolation is sometimes possible, without preliminary concentration (Hinshaw, Webster & Turner 1979). Transmission among ducks apparently takes place by the faecal-oral or faecal-cloacal routes. Lake waters and bird faeces are generally acid, but many wild duck isolates are relatively acid-resistant. This property is dependent on the haemagglutinin component being in the stable uncleaved (non-infectious) state. Uncleaved haemagglutinins are associated with a restricted ability to replicate in different cell types and with low virulence (see Alexander 1982b; Scholtissek 1986).

A recent extensive outbreak of avian influenza (subtype H5N2) in the eastern USA eventually resulted in the destruction of 17 million birds and expenditure of over $60 million on control measures (United States Department of Agriculture Mar., June, Dec. 1984, Mar. 1985). A virus (H5N2) of low virulence, probably originating in waterfowl, was isolated from chickens in April 1983 and had become highly virulent by October 1983 (Kawaoka & Webster 1985). The conversion was almost certainly due to accumulation of point mutations and not to genetic recombination; sequencing studies showed that there were four amino-acid differences between haemagglutinins of the isolates of low and high virulence and, in recombination studies, only a single point mutation at a critical site in the haemagglutinin was sufficient to convert the virus of low virulence to a fully virulent form; all the other genes necessary for virulence were present at the beginning (Webster, Kawaoka & Bean 1986). After emphasizing the importance of such mutations it was concluded that 'a practical step which can be taken to minimize the likelihood of another AI (avian influenza) epidemic is to prevent direct or indirect contact between aquatic birds and domestic poultry' (Bean, Kawaoka, Wood, Pearson & Webster 1985; Webster, Kawaoka & Bean 1985, 1986). This is a lesson likely to be increasingly recognized.

An important question, which has been partially answered in the last few years, is whether avian and mammalian influenza viruses can infect mammals and birds respectively. The answer to the first part of the question would need to be 'yes', if it is to be postulated that the enormous pool of influenza virus genes in wildfowl constitutes a potential risk for man and domestic mammals. Tentative followed by conclusive evidence for transfer of H1N1 virus from pigs to turkeys and man has now been obtained, and there is genetic and antigenic information showing that pigs sometimes acquire infection from avian sources (Hinshaw, Alexander, Amard *et al.* 1984).

Possibly the most spectacular instance of transfer of influenza virus from an avian to a mammalian host was the occurrence in 1979/81 of an H7/N7 strain in harbour seals (*Phoca vitulina*) off Cape Cod, Mass., USA. About 500 of these animals (*c.* 20% of the population) died with lesions of acute haemorrhagic pneumonia, in which virus was present to high titre. A similar but less severe disease was reproduced in experimental seals and the virus also replicated silently in ferrets, cats and pigs, and in the respiratory tract of avian species; it caused conjunctivitis in man. There was no serological evidence for infection of seals in other areas, which suggests that infection was of recent origin. Antigenic and biochemical (RNA segment reassociation) assays showed that the seal virus was closely related to H7/N7 virus isolated from wild ducks in Iceland in 1980. In other words similar viruses were circulating in waterfowl at the time of the outbreak in seals (Webster, Hinshaw, Bean & van Wyke 1981).

Viruses having comparative scientific interest

It may seem surprising to choose the viruses of foot-and-mouth disease (FMD) and vesicular exanthema (VE) as examples of this group, especially FMD virus which was long ago suspected to be harboured in African buffalo populations (DeKock 1946; DuToit 1947). Later research has nevertheless shown that the buffalo contribution to FMD epidemiology is probably less than was once believed; VE is highly unusual in having pelagic mammals as the probable reservoir for a disease of domestic animals. Both viruses throw some light on the manner in which viruses pathogenic to livestock may have arisen (see also Hammond & Branagan 1973).

Foot-and-mouth disease

A very large number of species, in many Orders and Families, are susceptible to infection by FMD viruses, but there is little evidence that they have had epidemiological importance outside Africa (Brooksby 1969; Hedger 1981). In the south of that continent there was by the 1940s increasing but largely anecedotal evidence that the buffalo (*Syncerus caffer*) was particularly important as a source of virus for cattle (see, for example, Thomas & Neitz 1933; DeKock 1946).

The development of serological techniques for typing FMD viruses led to a clearer definition of the problem. Generally speaking, buffalo and other game animals were found to be infected with the same SAT types of virus that, simultaneously or shortly afterwards, caused disease in domestic cattle. In some instances these types had not been recorded for many years or had never been encountered before in cattle of the country concerned. In addition a prolonged virus carrier state (up to 2.5 years) in some recovered cattle was recognized, and techniques were devised for the detection of virus in oesophageal and pharyngeal secretions (Hedger 1976) and for large-scale antibody surveys.

The work of Hedger and co-workers from 1969 onwards (Hedger 1976) established that buffaloes over the whole of their range in eastern and southern Africa were very frequently long-term carriers of all three SAT types of FMD virus and that the viruses were fully pathogenic for cattle. Clinical disease was never, or rarely, observed in buffaloes, although antibody studies showed virtually universal infection with all three SAT serotypes by the age of 2–3 years, often perhaps under cover of passive (colostral) antibody. Furthermore, small populations—57 in the case of an island in Lake Kariba—were capable of the indefinite maintenance of FMD virus (up to 24 years) without transfer to other susceptible species occurring (see Condy, Hedger, Hamblin & Barnett 1985).

It was for a long time impossible to demonstrate experimental infection of cattle by close, long-term contact with carrier buffaloes. Nevertheless, in Zimbabwe large areas of farmland have recently been completely cleared of buffalo and fenced against re-invasion because of the imputed risks to cattle (Bengis, Thomson, Hedger, de Vos & Pini 1986). The first successful transfer of FMD virus from carrier buffaloes to cattle took place after two years' contact on a peninsula in Lake Kariba—virus excretion from the carriers possibly being precipitated by the stress of a 70-mile road journey (Hedger & Condy 1985). The many failures to demonstrate transmission from buffaloes to cattle (Hedger 1976; E. C. Anderson, Doughty, Anderson & Paling 1979; Gainaru, Thomson, Bengis, Esterhuysen, Bruce & Pini 1986), or even to their own calves over a period of 15 months after birth, lead to the conclusion that such events are relatively rare and that the main method of transmission, whether to their own kind or other species such as impala or cattle, is from acutely infected buffaloes excreting large amounts of virus, possibly in respiratory aerosols which reach recipients in very close proximity (Bengis *et al.* 1986; Gainaru *et al.* 1986). The application of T1 ribonuclease oligonucleotide mapping to SAT 1 and SAT 2 isolates, recovered from buffalo and cattle in the same areas of Kenya at about the same time showed that viruses circulating in cattle were genomically distinct from those in buffaloes (E. C. Anderson 1986).

So far as other wildlife species in Africa are concerned, the virus carrier state has not so far been reported in them, but outbreaks of clinically recognizable disease, often confirmed virologically, have been recorded since the 1930s in species such as kudu and, since the 1940s, in impala, particularly in and near the Kruger National Park (KNP) in South Africa. Species that develop clinical FMD may be severely lame and suffer loss of condition as a result of pain in the prehension and mastication of food. The mortality rate is unknown, but it is reasonable to suppose that many of those affected fall easily to predators.

Thomson, Bengis, Esterhuysen & Pini (1984) listed 14 outbreaks in impala in the KNP between 1967 and 1983 of which 12 were due to types SAT 1 and SAT 2, two being untyped. All the extensive and more prolonged outbreaks occurred before December 1977 and in the dry season or during winter, i.e. at times of the year when impala were most likely to congregate around water points where they could make close contact with carrier buffalo. The lack of severe outbreaks in recent years may have been due to increasing population densities of both buffalo and impala, which have virtually doubled from 1967 and 1978 respectively; possibly increasing density is associated with enzootic infection and populations of impala that are partly immune and no longer capable of supporting extensive epizootics. It was concluded (Thomson *et al.* 1984) that the ecology of the disease in and around the KNP needed further elucidation.

Vesicular exanthema (VE)

This vesicular disease, affecting pigs only, was first seen in California in 1932, and between that year and 1936 caused 10 distinct outbreaks, all of which were on the west coast of the USA; they were controlled by slaughter or quarantine and successfully eradicated. Between 1939 and 1953 the disease spread from California to all the swine-raising areas of the USA and was brought under control and eradicated only in 1956 by a Federal programme enforcing the heat-treatment of garbage for pig feeding. The virus was important above all for the resemblance of the lesions caused to those of FMD. Its origins were not determined; but it was known to be antigenically diverse, 13 distinct serotypes being recognized by virus neutralization tests (Bankowski 1965).

In 1972 a calicivirus, physicochemically identical to that of VE (Burroughs, Doel & Brown 1978), was isolated from an aborting sealion on an island off the California coast (San Miguel); this animal (*Zalophus californianus*) also had vesicular lesions on the flippers. The San Miguel sealion virus (SMSV-1) was serologically distinct from any strain of VE virus and 12 further distinct serotypes had been isolated from marine mammals off the Pacific coast of North America by 1984; seven original isolates came from Californian sealions, five from northern fur seals (*Callorhinus ursinus*), and one from a northern elephant seal (*Mirounga angustirostris*) (Smith 1983, 1986). In addition, isolates were obtained from the opal-eye fish (*Girella nigricans*), which is presently regarded as a probable primary source.

The pinniped caliciviruses are transmissible to pigs, in which they sometimes cause lesions more severe than those produced by any of the original 13 serotypes of swine origin; natural infection has also been recorded in man and antibodies are present in feral swine, donkeys, buffalo and sheep, as well as foxes, on offshore islands. They are also present in five species of whale (grey, sei, sperm, fin and bowhead).

The mechanisms of spread from marine to terrestrial mammals are suspected to include infection through skin or mucosal abrasions, for example on contaminated beaches. Mink are infected when uncooked seal meat is fed to them. It is suspected that a similar mechanism was responsible for the original outbreaks in swine, as discarded remnants were known to have been given to the pigs. Experimentally, monkeys (*Cercopithecus aethiops*) are susceptible and a calicivirus occurred naturally in vesicular lesions of primates in the San Diego zoo (Smith 1983). The lesions in northern fur seals are located on the hairless parts of the flippers, and vesicles develop essentially as in FMD. In the 25,000 or so fur seals that are 'harvested' annually on St Paul Island, Alaska, the frequency of visible lesions was estimated to be only 0.1%, but figures of up to 3% have been reported elsewhere. Experimental lesions in susceptible seals were mild and healed without rupture (Gelberg, Dietrich & Lewis 1982).

It is not surprising that these investigations, which appear to have established the existence of widespread calicivirus infections in pinnipeds and possibly whales in the west Pacific, have been primarily concerned with the dangers to domestic species. The sources of finance for much of the work and the preoccupations of the investigators have, however, resulted in a dearth of information on the effects of the agents on their original hosts. It remains to be proved whether caliciviruses cause abortions and neonatal deaths in sealions and have appreciable effects on pinniped populations.

Viruses important in animal welfare and conservation

It is often apparently forgotten that infectious diseases generally have serious implications for animal welfare. I have not attempted to introduce rabies virus here as a major example, but R. M. Anderson (1986) pointed out that a minimum of 1.25 million foxes are reported as killed in Europe annually in efforts to control rabies, which affects only 1–4 persons every year. And how many tears were shed for the *c.* 13,000 foxes confirmed to WHO as rabid in 1985? The concept of what might be called 'taxonomic weighting' in the conservation ethic was described thus by Dobson & May (1986) '. . . for most people, any member of our own multitudinous species deserves more concern than the last wild members of other great ape species; large animals excite more compassion than small ones; mammals outrank other vertebrates (except some birds) and vertebrates outrank invertebrates.'

Anybody who has seen advanced cases of malignant catarrhal fever (MCF) in ruminants—and virtually all cases are fatal—will realise that the disease has welfare as well as conservational implications. Hence MCF is the final example given in this paper. It could also be used to illustrate how information acquired over the last 50–60 years in Europe and Africa primarily for the benefit of cattle has facilitated, if not made possible, work on species in zoos.

Malignant catarrhal fever

This disease, caused by one of a number of closely-related, lymphotropic herpesviruses, is becoming increasingly recognized as an important source of disease and mortality in farmed deer and in ruminants in zoological collections. Some of the viruses have natural reservoirs in alcelaphine and hippotragine antelopes (wildebeest, topi, sable antelope, oryx, etc.) in Africa and, because they cause persistent inapparent and vertically transmitted infections in these species, have been transferred unwittingly to the majority of zoo populations throughout the world and also to many 'game ranches' or wildlife sanctuaries in North America. They are also probably found in wild caprine species and domestic sheep worldwide. Outside Africa, sheep are the source of the great majority of outbreaks of MCF in cattle and deer (Plowright 1981, 1986). Recent serological work in North America shows that antigenically related viruses are widespread in wild caprine species such as the Barbary sheep, *Ammotragus lervia* (Heuschele, Fletcher, Oosterhuis, Janssen & Robinson 1984). However, none of the putative viruses in Caprinae has yet been cultured, as such, or characterized (see Plowright 1986).

Both the alcelaphine and caprine agents are transferred from reservoir to susceptible 'indicator' hosts by very young, even neonatal, animals. The alcelaphine (wildebeest) agents are excreted in infectious form in the nasal and ocular secretions of the young, up to the time when local antibodies appear. Hence the disease tends to be acquired seasonally by indicator hosts. The list of indicator species, which has gradually lengthened in the last decade or so, includes the Asiatic wild cattle (gaur and banteng), the European and American bison, eland, kudu, sitatunga, nilgai and many Cervidae, including red deer (*Cervus elaphus*) and Père David deer (*Elaphurus davidianus*). Infection normally requires close contact. Sporadic cases are frequent, but incidents in zoological collections often affect relatively large numbers of valuable or even threatened species. One reason for this is that the danger and source are often recognized only after several animals have been exposed and after the disease has developed in one or more of them. Moreover, the prolonged incubation period of four to 12 weeks or more increases the risk of multiple cases, occurring as the result of prolonged exposure to unsuspected donors.

In relation to welfare, it should be stated that the infection in indicator species is as lethal as it is harmless in the reservoirs and very few survive. Those that die quickly, within 48 h, often have severe diarrhoea or dysentery; those surviving longer, up to 10–14 days, frequently suffer from severe respiratory distress, stomatitis, corneal opacity and blindness, signs of central nervous system involvement, skin eruptions, and generalized lymphadenopathy. Affected cattle and deer are often so pathetic that they are destroyed on humanitarian grounds.

The potential, sometimes real, threat of MCF to survival of a rare species or of a species for which a captive breeding programme is in train, is illustrated by the figures for Père David deer—eight of a herd of 14 were lost at Whipsnade Park (Zoological Society of London) between April and July 1959 and the rest were killed (Huck, Shand, Allsop & Patterson 1961); 10 died of MCF in the San Diego Zoo between 1980 and 1984 and, furthermore, 11 banteng were lost between 1974 and 1982 (Heuschele 1982).

Fortunately the principles of control of MCF are relatively simple: indicator and reservoir species must be separated especially during the season when neonatal reservoir hosts are present to excrete infective virus. The necessity for isolation can now be precisely based, on serological confirmation of the presence or absence of lymphotropic herpesviruses in potential reservoir hosts. In Africa, the great danger is that increasing pressures on grazing will lead to calls for the removal—often extermination—of wildebeest from areas which they now share with pastoral peoples or ranches. Only the development of an effective vaccine for cattle could, in the long term, moderate such demands.

Discussion

As with virtually all wildlife diseases, those caused by viruses have usually been selected for investigation because they are, or are thought to be, transmissible to man, domestic animals, or both. This has, understandably but unfortunately, led to a neglect of diseases affecting wildlife populations *per se* and sometimes, undoubtedly, to draconian measures against wildlife based on less than adequate information or consideration of alternatives. Rabies has been mentioned in this context (R. M. Anderson 1986), and also foot-and-mouth disease (Bengis *et al.* 1986). It is self-evident that such measures are impracticable for the control of certain infections, such as those due to paramyxoviruses and myxoviruses in migratory waterfowl.

It is regrettable that the presumed reaction of the authorities responsible for animal or human health to the existence of important infectious diseases in wildlife (e.g. rinderpest, FMD), has led to a tendency for some biologists to conceal or discourage investigation of the evidence. Some also have clung to the belief that disease is not a significant factor in the control or conservatin of wildlife populations. Thus, Eltringham (1979) stated 'Disease is a natural factor in wild animals and rarely requires management attention'. Fortunately, a more enlightened attitude is now spreading, particularly as a result of the publications of R. M. Anderson and R. M. May, who have stressed the importance of disease, predators and resource limitation as co-factors in population biology (R. M. Anderson & May 1979, 1982; May & Anderson 1979). It is noteworthy that some recent research, as for example with FMD and rinderpest in Africa, does not strengthen the case for action against wildlife so much as point to the need for continuing studies before taking drastic action.

Generally speaking, continuous monitoring of infectious diseases in wildlife is essential, if catastrophes such as the one that overtook the black-footed ferret are to be avoided. The monitoring should comprise much more than periodic counts of 'natural' deaths or even observations of population trends. Serological surveys with agent recovery are essential and feasible measures; they should be statistically valid, taking into account geographically distinct populations, the various susceptible species, and different age groups.

It would also be desirable to begin to apply the techniques of genetic manipulation to viruses in the search for live attenuated strains that could be used as vaccines for wildlife and other species. Such vaccines should be safe, capable of spreading naturally (e.g by the oral route or by contact) and capable of conferring long-term immunity. They would help to invalidate the complaint that there is little that can be done to control infectious diseases in wild animals.

References

Alexander, D.J. (1982a). Avian paramyxoviruses—other than Newcastle disease virus. *Wld's poult. Sci. J.* **38**: 97–104.

Alexander, D.J. (1982b). Avian influenza—recent developments. *Vet. Bull.* **52**: 341–59.

Alexander, D.J., Allan, W.H., Parsons, G. & Collins, M.S. (1982). Identification of paramyxoviruses isolated from birds dying in quarantine in Great Britain during 1980 to 1981. *Vet. Rec.* **111**: 571–4.

Alexander, D.J., Hinshaw, V.S., Collins, M.S. & Yamane, N. (1985). Characterisation of viruses which represent further distinct serotypes (PMV-8 and PMV-9) of avian paramyxoviruses. *Archs Virol.* **78**: 29–36.

Alexander, D.J., Lister, S.A. & Wilson, G.W.C. (1986). Avian paramyxovirus type 1 infections of racing pigeons: 5. Continued spread in 1984. *Vet. Rec.* **118**: 424–7.

Alexander, D.J., Russell, P.H. & Collins, M.S. (1984). Paramyxovirus type 1 infections of racing pigeons. 1. Characterisation of isolated viruses. *Vet. Rec.* **114**: 444–6.

Alexander, D.J., Wilson, G.W.C., Russell, P.H., Lister, S.A. & Parsons, G. (1985). Newcastle disease outbreaks in fowl in Great Britain during 1984. *Vet. Rec.* **117**: 429–34.

Anderson, E.C. (1986). Potential for the transmission of foot-and-mouth disease in the African buffalo (*Syncerus caffer*) to cattle. *Res. vet. Sci.* **40**: 278–80.

Anderson, E.C., Doughty, W.J., Anderson, J. & Paling, R. (1979). The pathogenesis of foot-and-mouth disease in the African buffalo (*Syncerus caffer*) and the role of this species in the epidemiology of the disease in Kenya. *J. comp. Path.* **89**: 541–9.

Anderson, R.M. (1986). Vaccination of wildlife reservoirs. *Nature, Lond.* **322**: 304–5.

Anderson, R.M. & May, R.M. (1979). Population biology of infectious diseases: part 1. *Nature, Lond.* **280**: 361–7.

Anderson, R.M. & May, R.M. (Eds) (1982). *Population biology of infectious diseases*. Springer-Verlag, Berlin. (*Life Sci. Res. Rep.* No. 25).

Anderson, R.M. & May, R.M. (1986). The invasion, persistence and spread of infectious diseases within animal and plant communities. *Phil. Trans. R. Soc.* (B) **314**: 533–70.

Bankowski, R.A. (1965). Vesicular exanthema. *Adv. vet. Sci.* **10**: 23–64.

Barnes, A.M. (1982). Surveillance and control of bubonic plague in the United States. *Symp. zool. Soc. Lond.* No. 50: 237–70.

Baxby, D. (1977a). Poxvirus hosts and reservoirs. *Archs Virol.* **55**: 167–79.

Baxby, D. (1977b). Is cowpox misnamed? A review of 10 human cases. *Br. med. J.* **1977** (1): 1379–81.

Baxby, D., Ashton, D.G., Jones, D.M. & Thomsett, L.R. (1982). An outbreak of cowpox in captive cheetahs: virological and epidemiological studies. *J. Hyg., Camb.* **89**: 365–72.

Baxby, D., Ashton, D.G., Jones, D.M., Thomsett, L.R. & Denham, E.M. (1979). Cowpox virus infection in unusual hosts. *Vet. Rec.* **109**: 175.

Baxby, D. & Ghaboosi, B. (1977). Laboratory characteristics of pox-viruses isolated from captive elephants in Germany. *J. gen. Virol.* **37**: 407–14.

Bean, W.J., Kawaoka, J.M., Wood, J.M., Pearson, J.E. & Webster, R.G. (1985). Characterisation of virulent and avirulent A/Chicken/Pennsylvania/83 influenza viruses: potential role of defective interfering RNAs in nature. *Virology* **54**: 151–60.

Becker, W.B. (1966). The isolation and classification of Tern virus: influenza virus A/Tern/South Africa/1961. *J. Hyg., Camb.* **64**: 309–20.

Bengis, R.G., Thomson, G.R., Hedger, R.S., de Vos, V. & Pini, A. (1986). Foot-and-mouth disease in African buffalo (*Syncerus caffer*). 1. Carriers as a source of infection for cattle, *Onderstepoort J. vet. Res.* **53**: 69–73.

Bennett, M., Baxby, D., Gaskell, R.M., Gaskell, C.J. & Kelly, D.F. (1985). The laboratory diagnosis of Orthopoxvirus infection in the domestic cat. *J. small anim. Pract.* **26**: 653–61.

Bennett, M., Gaskell, R.M., Gaskell, C.J., Baxby, D. & Gruffydd-Jones, T.J. (1986). Poxvirus infection in the domestic cat: some clinical and epidemiological observations. *Vet. Rec.* **118**: 387–90.

Berry, H.H. (1981). Abnormal levels of disease and predation as limiting factors for wildebeest in the Etosha National Park. *Madoqua* **12**: 242–53.

Brooksby, J.B. (1969). Wild animals and the epizootiology of foot-and-mouth disease. *Symp. zool. Soc. Lond.* No. 24: 1–11.

Burroughs, N., Doel, T. & Brown, F. (1978). Relationship of San Miguel sea lion virus to other members of the calicivirus group. *Intervirology* **10:** 51–9.

Carpenter, J.W., Appel, M.J.G., Erickson, R.C. & Novilla, M.N. (1976). Fatal vaccine-induced canine distemper virus infection in black-footed ferrets. *J. Am. vet. med. Ass.* **169:** 961–4.

Condy, J.B., Hedger, R.S., Hamblin, C. & Barnett, I.T.R. (1985). The duration of the FMDV carrier state in African buffalo: 1) in the individual animal and 2) in a freeliving herd. *Comp. Immun. Microbiol. inf. Dis.* **8:** 259–65.

DeKock, G. (1946). Problems of game conservation in Southern Africa. *S. Afr. J. Sci.* **42:** 162–71.

Dobson, A.P. & May, R.M. (1986). Disease and conservation. In *Conservation biology: the science of scarcity and diversity*. (Ed. Soulé, M.E.). Blackwell, Oxford.

Dunkin, G.W. & Laidlaw, P.P. (1926). Studies on dog distemper 1. Dog distemper in the ferret. *J. comp. Path. Ther.* **39:** 201–12.

Du Toit, P.J. (1947). Game in relation to animal diseases. *J. S. Afr. vet. med. Ass.* **18:** 59–66.

Eltringham, S.K. (1979). *The ecology and conservation of large African mammals*. The Macmillan Press, Ltd., London & Basingstoke.

Fenner, F. (1983). Biological control, as exemplified by smallpox eradication and myxomatosis. *Proc. R. Soc.* (B) **218:** 259–85.

Frame, L.H., Malcolm, J.R., Frame, G.W. & Lawick, H.van (1979). Soical organization of African wild dogs (*Lycaon pictus*) on the Serengeti plains, Tanzania 1967–1978. *Z. Tierpsychol.* **50:** 225–49.

Friend, M. (1976). Wildlife diseases: philosophical considerations. In *Wildlife diseases*: 7–18. (Ed. Page, L.A.). Plenum Press, New York and London.

Gainaru, M.D., Thomson, G.R., Bengis, R.G., Esterhuysen, J.J., Bruce, W. & Pini, A. (1986). Foot-and-mouth disease and the African buffalo (*Syncerus caffer*). II. Virus excretion and transmission during acute infection. *Onderstepoort J. vet. Res.* **53:** 75–85.

Gaskell, R.M., Gaskell, C.J., Evans, R.J., Dennis, P.E., Bennet, A.M., Undall, N.D., Voyle, C. & Hill, T.J. (1983). Natural and experimental pox virus infection in the domestic cat. *Vet. Rec.* **112:** 164–71.

Gelberg, H.B., Dietrich, R.A. & Lewis, R.M. (1982). Vesicular exanthema of swine and San Miguel sea lion virus: experimental and field studies in otarid seals, feeding trials in swine. *Vet. Path.* **19:** 413–23.

Hammond, J.A. & Branagan, D. (1973). The disease factor in plans for the domestication of wild ruminants in Africa. *Vet. Rec.* **92:** 367–9.

Hanson, R.P. (1976). Avian reservoirs of Newcastle disease. In *Wildlife diseases*: 185–95. (Ed. Page, L.A.). Plenum Press, New York and London.

Hedger, R.S. (1976). Foot-and-mouth disease in wildlife with particular reference to the African buffalo (*Syncerus caffer*). In *Wildlife diseases*: 235–44. (Ed. Page, L.A.). Plenum Press, New York and London.

Hedger, R.S. (1981). Foot-and-mouth disease. In *Infectious diseases of wild mammals* (2nd edn): 87–96. (Eds Davis, J.W., Karstad, L.H. & Trainer, D.O.). Iowa State University Press, Ames.

Hedger, R.S. & Condy, J.B. (1985). Transmission of foot-and-mouth disease from African buffalo virus carriers to bovines. *Vet. Rec.* **117**: 205.

Heuschele, W.P. (1982). Malignant catarrhal fever in wild ruminants—a review and current status report. *Proc. a. Meet. U.S. anim. Hlth Ass.* **86**: 557–70.

Heuschele, W.P., Fletcher, H.R., Oosterhuis, J., Janssen, D. & Robinson, P.T. (1984). Epidemiological aspects of malignant catarrhal fever in the USA. *Proc. a. Meet. U.S. anim. Hlth Ass.* **88**: 640–51.

Hinshaw, V.S., Alexander, D.J., Amard, M. *et al.* (1984). Antigenic comparisons of swine-influenza-like H1N1 isolates from pigs, birds and humans: an international collaborative study. *Bull. Wld. Hlth Org.* **62**: 871–8.

Hinshaw, V.S., Webster, R.G. & Turner, B. (1979). Water-borne transmission of influenza A viruses. *Intervirology* **11**: 66–8.

Hinshaw, V.S., Webster, R.G. & Turner, B. (1980). The perpetuation of orthomyxoviruses and paramyxoviruses in Canadian waterfowl. *Can. J. Microbiol.* **26**: 622–9.

Hoare, C.M., Gruffydd-Jones, T.J., Bennett, M., Gaskell, R.M. & Baxby, D. (1984). Cowpox in cats. *Vet. Rec.* **114**: 22.

Huck, R.A., Shand, A., Allsop, P.J. & Patterson, A.B. (1961). Malignant catarrh of deer. *Vet. Rec.* **73**: 457–65.

Kaplan, C., Healing, T.D., Evans, N., Healing, L. & Prior, A. (1980). Evidence of infection by viruses in small British field rodents. *J. Hyg., Camb.* **84**: 285–94.

Karstad, L. (1981). Miscellaneous viral infections. In *Infectious diseases of wild animals* (2nd edn.): 202–11. (Eds Davis, J.W., Karstad, L.H. & Trainer, D.O.). Iowa State University Press, Ames.

Kawaoka, Y. & Webster. R. (1985). Evolution of the A/Chicken/Pennsylvania/83 (H5N2) influenza virus. *Virology* **146**: 130–7.

Lister, S.A., Alexander, D.J. & Hogg, R.A. (1986). Evidence for the presence of avian paramyxovirus type 1 in feral pigeons in England and Wales. *Vet. Rec.* **118**: 476–9.

Marennikova, S.S., Ladnyj, I.D., Ogorodnikova, Z.I., Shelukhina, E.M. & Maltseva, N.N. (1978). Identification and study of a poxvirus isolated from wild rodents in Turkmenia. *Archs Virol.* **56**: 7–14.

Marennikova, S.S., Maltseva, N.N., Korneeva, V.I. & Garanina, N.M. (1977). Outbreak of pox disease among Carnivora (Felidae) and Edentata. *J. inf. Dis.* **135**: 358–66.

Martin, W.B., Scott, F.M.M., Lauder, I.M. & Nash, A. (1984). Poxvirus infections of cats. *Vet. Rec.* **115**: 36.

Martland, M.F., Fowler, S., Poulton, G.J. & Baxby, D. (1983). Poxvirus infection of a domestic cat. *Vet. Rec.* **112**: 171–2.

Martland, M.F., Poulton, G.J. & Done, R.A. (1985). Three cases of cowpox infection of domestic cats. *Vet. Rec.* **117**: 231–3.

May, R.M. (1986). The cautionary tale of the black-footed ferret. *Nature, Lond.* **320**: 13–14.

May, R.M. & Anderson, R.M. (1979). Population biology of infectious diseases: part 3. *Nature, Lond.* **280**: 455–61.

Moehlman, P.D. (1983). Socioecology of silverbacked and golden jackals (*Canis mesomelas* and *Canis aureus*). *Spec. Publs Am. Soc. Mammal.* No. 7: 423–53.

Neitz, W.O. (1967). The epidemiological pattern of viral, protophytal and protozoal zoonoses in relation to game preservation in South Africa. *J. S. Afr. vet. med. Ass.* **38:** 129–41.

O'Brien, S.J., Roelke, M.E., Marker, L., Newman, A., Winkler, C.A., Meltzer, D., Colly, L., Evermann, J.F., Bush, M. & Wildt, D.E. (1985). Genetic basis for species vulnerability in the cheetah. *Science, N.Y.* **227:** 1428–34.

Pether, J.V.S., Trevains, P.H., Harrison, S.R.B., Baxby, D., Bennett, M. & Gibb, A.P.R. (1986). Cowpox from cat to man. *Lancet* **1986** (i): 38.

Plowright, W. (1981). Herpesviruses of wild ungulates including malignant catarrhal fever virus. In *Infectious diseases of wild mammals* (2nd edn): 126–46. (Eds Davis, J.W., Karstad, L.H. & Trainer, D.O.). Iowa State University Press, Ames.

Plowright, W. (1982). The effects of rinderpest and rinderpest control on wildlife in Africa. *Symp. zool. Soc. Lond.* No. 50: 1–28.

Plowright, W. (1986). Malignant catarrhal fever. *Rev. scient. tech. Off. int. Epizoot.* **5:** 455–68.

Plowright, W. & McCulloch, B. (1967). Investigations on the incidence of rinderpest virus infection in game animals of N. Tanganyika and S. Kenya 1960/63. *J. Hyg., Camb.* **65:** 343–58.

Reid, H.W. (1984). Epidemiology of louping ill. In *Vectors in virus biology.* (Eds Mayo, M.A. & Harrop, K.H.). Academic Press, London.

Reynolds, J.C. (1985). Details of the geographic replacement of the red squirrel (*Sciurus vulgaris*) by the grey squirrel (*Sciurus carolinensis*) in eastern England. *J. anim. Ecol.* **54:** 149–62.

Ross, J. (1982). Myxomatosis: the natural evolution of the disease. *Symp. zool. Soc. Lond.* No. 50: 77–95.

Rossiter, P.B. (1986). Rinderpest antibodies in buffaloes in the Serengeti National Park and Masai Mara Game Reserve. *Newsl. vet. Specialist Group SSC/IUCN* No. 2: 7–9.

Rossiter, P.B., Jessett, D.M., Wafula, J.S., Karstad, L., Chema, S., Taylor, W.P., Rowe, L.W., Nyange, J.C., Otaru, M., Mumbala, M. & Scott, G.R. (1983). The re-emergence of rinderpest as a threat in East Africa, 1979–82. *Vet. Rec.* **113:** 459–61.

Rossiter, P.B., Taylor, W.P., Bwangamoi, B., Ngereza, A.R.H., Moorhouse, P.D.S., Haresnape, J.M., Wafula, J.S., Nyange, J.F.C. & Gumm, I.D. (1987). Continuing presence of rinderpest virus as a threat in East Africa, 1983–85. *Vet. Rec.* **120:** 59–62.

Sands, J.J., Scott, A.C. & Harkness, J.W. (1984). Isolation in cell culture of a pox virus from the red squirrel (*Sciurus vulgaris*). *Vet. Rec.* **114:** 117–18.

Schiemann, B., Plowright, W. & Jessett, D.M. (1971). Allerton-type herpesvirus as a cause of lesions of the alimentary tract in a severe disease of Tanzanian buffaloes (*Syncerus caffer*). *Vet. Rec.* **89:** 17–22.

Scholtissek, Ch. (1986). Stability of infectious influenza A viruses to treatment at low pH and heating. *Archs Virol.* **85:** 1–11.

Scott, A.C., Keymer, I.F. & Labram, J. (1981). Parapoxvirus infection of the red squirrel (*Sciurus vulgaris*). *Vet. Rec.* **109:** 202.

Scott, G.R. (1981). Rinderpest. In *Infectious diseases of wild mammals* (2nd edn): 18–30. (Eds Davis, J.W., Karstad, L.H. & Trainer, D.O.). Iowa State University Press, Ames.

Senne, D.A., Pearson, J.E., Miller, L.D. & Gustafson, G.A. (1983). Virus isolations from pet birds submitted for importation into the United States. *Avian Dis.* **27:** 731–44.

Sinclair, A.R.E. (1977). *The African buffalo: a study of resource limitation of populations.* University of Chicago Press, Chicago and London.

Sinclair, A.R.E. & Norton-Griffiths, M. (1979). *Serengeti: dynamics of an ecosystem.* University of Chicago Press, Chicago and London.

Slemons, R.D. & Easterday, B.C. (1976). The natural history of Type-A influenza viruses and wild waterfowl. In *Wildlife diseases*: 215-24. (Ed. Page, L.A.). Plenum Press, New York and London.

Smith, A.W. (1983). Focus on caliciviral disease. *Foreign Anim. Dis. Reps U.S. Dep. Agric. Anim. Pl. Hlth Insp. Serv. vet. Serv.* No. 11–3: 8–16.

Smith, A.W. (1986). Caliciviral disease update. *Foreign Anim. Dis. Reps U.S. Dep. Agric. Anim. Pl. Hlth Insp. Serv. vet. Serv.* No. 14–3: 5–6.

Spinage, C.A. (1962). Rinderpest and faunal distribution patterns. *Afr. Wild Life* **16:** 55–60.

Talbot, L.M. & Talbot, M.H. (1961). Preliminary observations on the population dynamics of the wildebeest in Narok District, Kenya. *E. Afr. agric. for. J.* **28:** 58–60.

Taylor, W.P. & Watson, R.M. (1967). Studies on the epizootiology of rinderpest in blue wildebeest and other game species of northern Tanzania and southern Kenya, 1965–7. *J. Hyg., Camb.* **65:** 537–45.

Thomas, A.D. & Neitz, W.O. (1933). The importance of disease in wild animals. *S. Afr. J. Sci.* **30:** 419–25.

Thomsett, L.R., Baxby, D. & Denham, E.M.H. (1977). Cowpox in the domestic cat. *Vet. Rec.* **103:** 567.

Thomson, G.R., Bengis, R.G., Esterhuysen, J.J. & Pini, A. (1984). Maintenance mechanisms for foot-and-mouth disease virus in the Kruger National Park and potential avenues for its escape into domestic animal populations. *Proc. Wld Congr. Dis. Cattle* **13**(1): 33–8.

Thorne, E.T. (1986). Diseases strike site of only known black footed ferret population. *Newsl. Vet. Specialist Group SSC/IUCN* No. 2: 15–17.

United States Department of Agriculture (1983–1985). *Foreign animal disease reports. United States Department of Agriculture, Animal and Plant Health Inspection Service Veterinary Services.* Published quarterly.

Vizoso, A.D. (1968). A red squirrel disease. *Symp. zool. Soc. Lond.* No. 24: 29–38.

Webster, R.G., Hinshaw, V.S., Bean, W.J. & van Wyke, K.L. (1981). Characterisation of an influenza A virus from seals. *Virology* **113:** 712–24.

Webster, R.G., Kawaoka, Y. & Bean, W.J. (1985). Avian influenza update. *Foreign Anim. Dis. Reps U.S. Dep. Agric. Anim. Pl. Hlth Insp. Serv. vet. Serv.* No. 13–3: 2–5.

Webster, R.G., Kawaoka, Y. & Bean, W.J. (1986). Molecular changes in A/Chicken/Pennsylvania/83 (H5N2) influenza virus associated with acquisition of virulence. *Virology* 149: 165–73.

Willemse, A. & Egberink, H.F. (1985). Transmission of cowpox virus infection from domestic cat to man. *Lancet* **1985**(i): 1515.

Yuill, T.M. (1981). Myxomatosis and fibromatosis. In *Infectious diseases of wild mammals* (2nd edn): 154–77. (Eds Davis, J.W., Karstad, L.H. & Trainer, D.O.). Iowa State University Press, Ames.

Index